THE
SIMPLEST-CASE
SCENARIO

How the Universe May Be
Very Different From
What We Think It Is

KARL CORYAT

To all the things that have died, died.
To all the things that have died, died.
To all the things that have died, died.
To all the things that have died, died.
They're all our friends, and they've died.
—JIM CARROLL (PARAPHRASED)

ISBN-13: 978-1-5377-4592-3
ISBN-10: 1-5377-4592-1

Printed in the United States of America.
First printing.

 **NULL SET PRESS**

Alameda, California

CONTENTS

CONTENTS

"The stream of knowledge is heading toward a non-mechanical reality; the universe begins to look more like a great thought than like a great machine."

—Sir James Jeans, English physicist in *The Mysterious Universe* (1930)

INTRODUCTION

First, a bit of science fiction:

The scene is Reykjavik, Iceland, December 2034. Scientists from around the world have gathered to address a concern in astrobiology that has grown into a full-on crisis. In the early 21st century, many Earth-like exoplanets — planets that orbit stars other than our own Sun — had been discovered and catalogued. In 2028, the Sagan satellite was launched to observe individual exoplanets and determine the composition of their atmospheres, using a sophisticated version of the old technique of spectroscopy. The focus was the search for alien life: If a planet hosted complex chemical processes, as happens with metabolism in living organisms, then this might be reflected in the chemical signature of that planet's atmosphere. For example, about one-fifth of Earth's atmosphere is oxygen, which is otherwise rarely found isolated in nature. Anyone measuring Earth's atmosphere from afar would know that there's something unusual about the chemistry on our planet.

The problem, in December 2034, was that no significantly unusual chemistry had been found, anywhere beyond Earth. From 2028–30, several thousand exoplanets had been analyzed for unusual chemical signatures. In 2031 the search was ramped up to thousands per day, then more. By 2034, over ten million exoplanets had been surveyed, and every last one was found to be ordinary: no free elements as reactive as oxygen, no compounds more complex than the simple amino acids and sugars that had been known for decades to exist on our own Solar System's comets and planets. In astrobiological terms, the universe appeared to be sterile. Dead. What's going on?

Beginning as early as the 1960s, astronomers listening on radio telescopes began wondering why, given a vast universe and billions of years for life elsewhere to evolve, there was no sign of intelligence out there. They even invented names, "the Great Silence" and "the Fermi paradox," for this non-finding. But that didn't slow the investigation: Scientific organizations were formed in the search for extraterrestrial intelligence (SETI). Earth's own radio transmissions began to dwindle in the face of fiber-optic, cellular, and satellite communications. Some asked whether intelligent life, and certainly non-intelligent life, would even use technology that employs radio frequencies, the only way they would be detectable via radio telescope. So, a broader search for anything life-like was needed. In our story, this gave birth to the Sagan satellite project. Surely if the search were not limited to technological intelligence — if it could detect *any* kind of otherwise-unexplainable, complex chemical processes, and surely if we could scan exoplanets by the millions — we would discover, at last, that we are not alone in the universe.

By 2034, even the most hopeful astrobiologists had reason to doubt this. At the Reykjavik Conference, most of the attendees still believed that complex chemistry must be out there, even if it was rare. But a growing, vocal faction raised another proposal, one that was difficult to accept: *We will never find life of any kind whatsoever, anywhere beyond our own Solar System.* No matter how far we search, even if the number of analyzed exoplanets approached every planet in the observable universe, complex chemistry of the kind seen in Earthly life would never be found. There is literally one and only one lineage of life in this universe, and it's here, on Earth.

. . .

The above story may actually play out. In 100 years, it may be routinely accepted that when it comes to life, Earth is the only game in town — schoolchildren may learn this just as commonly as they learn today that we live in the Milky Way galaxy. Now, whenever I suggest this to science-educated people, they invariably react, "But the overwhelming number of stars ..." or, "Just by chance alone ..." which is perfectly understandable. However, the ultimate nature of the universe may be very different from the scenario reflected in those objections. I'm talking about a radically different, and simpler, picture of how the world is fundamentally put together — and in that picture, the Fermi "paradox" becomes a prediction.

Over the past decade, I have become fascinated with the problems and paradoxes in contemporary physics and cosmology. There are more of these than most people realize, and I touch on many of them in this book. Some of the world's deepest-thinking scientists, beginning in the 20th century with John Archibald Wheeler and continuing today with people like Paul Davies and Carlo Rovelli, have been looking for outside-the-box solutions. For me, it began when I heard that the pioneering stem-cell biologist Robert Lanza had floated a radical proposal: Life produces the universe, rather than the other way around. Even though that statement may seem absurd on the surface, I was intrigued — so I looked up the article he had written for *American Scholar* magazine, where he introduced this concept he called "biocentrism." In it, he wrote of the conundrums involving the famous experiments of quantum mechanics, and also the peculiar laws of the universe, which, depending on how you look at them, may be considered "fine-tuned" for matter and life. Lanza considered these to be important clues. "When science tries to resolve its conflicts by adding and subtracting dimensions to the universe like houses on a Monopoly board," he wrote, "we need to look at our dogmas and recognize that the cracks in the system are just the points that let the light shine more directly on the mystery of life."

This book is an attempt to make sense of life and existence. Since childhood, I had suspected that there is something odd about our place in the universe — that the real world we see "out there" is, somehow, something other than what we think it is. Lanza's ideas spoke to me. I started following the threads that led him to his conclusion. In reading the work of scientists such as Wheeler and Rovelli, I noticed that the theme of information kept coming up again and again. However, aside from the quantitative context of how much information is contained in a message coming down a wire, physicists seemed to have differing ideas on what information actually *is*. John Wheeler went so far as to say that all matter in the universe is actually derived from bits of information (discussed in Chapter 1), a proposition that has tantalized scientists for decades. Could a deep understanding of information, where it's broken down to its essence and rebuilt, literally bit by bit, actually reveal the secrets of existence? Over the course of several years, I hit upon a synthesis of various theorists' approaches, a unique combination of ideas that might accurately describe the ultimate nature of the universe and how it came to be that way. I call it the simplest-case scenario.

I could have considered this merely a hare-brained idea, one of thousands put forward by self-described independent researchers. But then something unusual happened. I summarized my idea in an essay for a worldwide competition run by the Foundational Questions Institute and co-sponsored by *Scientific American* magazine, and ... I won a prize. To my astonishment, alongside physics professors, affiliates of the Perimeter Institute, and even George F.R. Ellis[1] — one of the world's top cosmologists, who co-authored a book with Stephen Hawking — there was me and my humble essay on the winners' page.

That was when I realized I might be on to something.

---

1 Ellis left a note in the comments section: "I really like the focus of this essay on informational mechanics as a 'generalization of quantum mechanics that embeds contextual data into descriptions of subsystem interactions' ... I also applaud your sensible take on quantum measurement."

Could information actually be the bottom layer of reality, the ultimate well from which all matter and energy springs? What exactly is information in this context? Is it possible that life does produce the universe, as Robert Lanza believes — and if so, what exactly is life? What exactly is the universe? Does consciousness fit into the picture of the ultimate nature of the world, as many mystics believe? John Wheeler (who also coined the terms *black hole* and *wormhole*) conjectured that the "stuff" of the universe reveals itself through physical acts of observation, a process he called *observer-participancy*. So, what exactly is observation? How did all of this get started, and why do we have physical evidence for the Big Bang? All of these questions need to be answered, if we are to gain any understanding of how matter, life, and existence ultimately derive from information.

You may ask yourself what is the point of all of this, since we already understand the universe extremely well. In a sense, certainly we do. We have discovered laws of physics that explain many things, and we are refining those understandings all the time. However, there remain questions behind our deeper understanding of how the world works, the most critical among them related to the experiments of quantum mechanics, discussed in Chapter 2. Science enthusiasts learn that the quantum world is weird and counterintuitive: Light can behave like both waves and particles under different conditions, things can seem to go backward in time and alter the past, and particles can seem to signal each other across vast distances much faster than light. The mysteries fascinated John Wheeler, and his conclusion was that our interpretation of these experiments is backward: By adopting the intuitive stance that *matter* is the ground of all existence, the information provided by quantum experiments becomes weird and counterintuitive. He suggested that instead, if we put information first, everything would fall into place. Wheeler died in 2008, and no one has figured out how to turn his conjecture into a coherent, self-consistent theory with the power to explain multiple things. The simplest-case scenario attempts to do that.

We will start off by asking whether information might be the ground of all being, and then we'll look at key quantum experiments to see if they can be understood in an information-grounded world. Chapters 3 and 4 take a close look at what information is and how it seems to operate. Chapter 5 returns to quantum mechanics and tries to formulate an interpretation that is based, entirely, on the kind of information discussed in the previous chapters. Chapter 6 looks at how a universe made of information could have evolved, and how it might have turned out differently. Chapter 7 explores why the universe appears to begin with considerable complexity, and why it appears to us to have been finely tuned. Chapter 8 tackles the question of why all observers see the same consistent universe, and Chapter 9 explains what this all means for sentient, intelligent humans.

Even though I have done everything I can to make this book as clear and understandable as possible, I have to warn you: The material is challenging. I am presenting a picture in which the world operates on the simplest possible principles, but that doesn't mean the path to understanding them is an easy one. This is because the Greek natural philosophers thousands of years ago may have made incorrect assumptions about the nature of things — assumptions that caused no concerns whatsoever before the 20th century, but which are now catching up to us in the form of unsolved problems, from cosmology to quantum entanglement to neuroscience. These assumptions were based on human intuition, but strangely, in following their lead we may have ended up assuming that our universe is much more complicated than it needs to be.

So, open your mind, and try to see things with fresh eyes. I hope that not only will you understand why I claim there is no alien life in our universe, you will glimpse the true nature of matter and energy, and see why the universe appears to be fine tuned — as well as how it could get that way without any kind of intelligent designer or creator calling the shots. You will gain a deep understanding of the thorny concept of time.

You will realize why the mind and the body are not separate things, i.e., that the mind–body problem is not a problem. You will gain an understanding of the profound unity of all living things. Finally, you will see how the universe might be *far, far* simpler than it is presently believed to be. This should delight fans of Occam's razor, the principle that the explanation with the fewest assumptions, complications, and conditions is the preferred one. A near-infinite (or infinite) universe of atoms and other particles colliding, for eons, will seem wildly extravagant. Meanwhile, the concept of a god pulling the strings of a puppet world will appear incredibly small-minded.

· · ·

In writing this book, I needed to represent mainstream science honestly. From Galileo rolling things down inclined surfaces to the latest data from CERN, there exists an experimental canon — the set of outcomes and findings of all experiments that have ever been performed — that forms the backbone of the global scientific enterprise. Professional scientists understand the importance of this, as they are trained to concoct theories that work within the experimental canon. Amateur theorists often do not; in order to believe their own conjectures, or to make their ideas seem plausible, they must deny parts of the experimental canon. But denying the experimental canon is scientific death, in the sense that if you do it, you will never be taken seriously by the scientific community. Therefore, when I refer to experiments that have been performed, I encourage you to check what I have claimed. I hope that my ideas will be judged by the standards of this experimental evidence, not by what seems intuitive or "common sense." Intuition has been well documented to lead human thought astray again and again throughout history, and I discuss some examples in this book. When considering the deepest questions about existence, we must resist the urge to let intuition be our guide.

We know how airplanes fly and how the Sun generates energy, but we do not know, for example, why certain experiments seem to change the past (as discussed in Chapter 2), what causes a spread-out wave of light to seemingly collapse into a tiny particle when it encounters a photographic plate, or what exactly makes the conditions of the early universe seem to be balanced just so, such that we can be here today to contemplate it all. The simplest-case scenario attempts to tie together these mysteries and many others, while staying true to all of the experimental evidence that's available. When I give my own opinions, I identify them as such. In the few instances where I may have an original insight to offer, I make it clear that the idea is mine, so that it can be given due scrutiny and skepticism.

I need to mention one more thing: *It may be that none of this is true.* The simplest-case scenario may not, in fact, describe how nature actually works. This should be a given for any speculative conjecture, but not all modern writers approach their pet theories in this manner. Amateur theorists (and some professionals!) can become intoxicated with their self-perceived brilliance. They write with religious assurance, as if their idea is the only one that could possibly be correct; their way is just the way it is. I mention this because my sentences may come across with a similar tone at times, particularly in the last few chapters. It's only because I wanted to avoid having to attach awkward and wordy qualifiers such as "perhaps it is the case that" and "imagine if it were true that" to sentences over and over. Instead, when I describe how the world is put together according to the simplest-case scenario, I do so straightforwardly — as if it's just the way things are. Please forgive me.

I've written blog posts and essays on this topic, I've made videos and I've answered comments, I've built a website and written a FAQ ... but it takes a short book to really explain this new way of thinking, and the manner in which it ties together many diverse realms of the human experience. Thank you for giving it a try.

TWO APPROACHES TO THE NATURE OF THINGS

L et's start with what we know. That seems like a simple enough place to begin, right?

It isn't. The question of what we know is fraught with difficulty, and it's the subject of controversy and debate going back hundreds and even thousands of years. We need to address it right here, in Chapter 1.

When I say "what we know," I'm not referring to what we learned in school. That's what we've been taught, and some of it might be wrong. The same can be said of human knowledge in general; it's frail and subject to error and revision. What do we really know about the world? When you get right down to it, are there things we can be absolutely certain of — things that are not, by any stretch of the imagination, a matter of human analysis and interpretation?

To strip this question down to its essence, let's imagine an experiment. In a sealed box, we place a radioactive sample, along with a Geiger counter tube. The rest of the detector, including the speaker, is outside the box. Then we sit and wait for something to happen. After a while, we hear the Geiger counter click.

What do we know happened?

Let's think this through. We heard a click. By that I mean, our brain became conscious of a sound, which presumably resulted from a sound wave impacting our eardrums. The sound wave presumably came from the Geiger counter's speaker, which presumably was driven by an amplified signal that began with an avalanche of electrons in the Geiger counter tube (this is how Geiger counters work). Presumably, that avalanche was caused by a particle entering the tube (that is what Geiger counters are made to detect), and since the box was sealed, we presume the particle came from the radioactive sample. The sample was uranium, so we presume it was an alpha particle consisting of two protons and two neutrons, since scientific knowledge tells us that these particles are given off by the nuclei of uranium atoms.

Observation: We heard a click. Conclusion: An alpha particle was emitted by our sample. Despite all of the presumably's in the last paragraph, it would be difficult to challenge the conclusion, as each link in the chain is well understood. (Even the first: We may not know *how* consciousness works, but it does seem to work!) Therefore, we are inclined to say, "We know that an alpha particle was emitted." And we feel highly justified in saying so.

But, what do we really know?

I claim that what we truly know about the situation is less than the paragraph would have us believe. *Something happened.* All other aspects of the description — the business about the composition of the particle, the electrons in the wires, even the when and where of the event — are components of a rich story that we arrive at by interpreting one piece of new information ("click!") within a picture that describes the experimental setup and how the various components work.

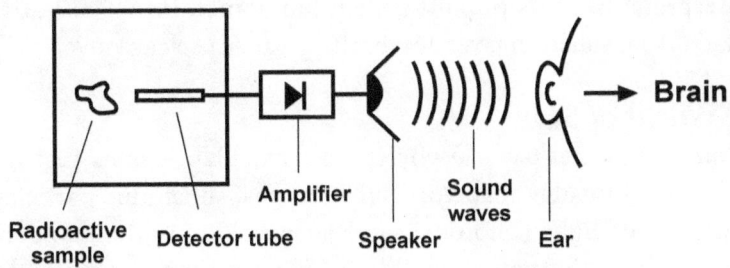

Fig. 1. Even a simple experimental setup can involve a complex chain of events.

Suppose you set up the experiment: You chose the location and time, you chose the sample, you chose a detector, and you sealed the box. You recorded when the click happened and interpreted the results with professional rigor. Still, the only thing you can know *for certain* about that particle and its history is: a click. Surely, a click sound registered in your consciousness. Perhaps that's all that happened; maybe you imagined it. But if similar events also occurred in the consciousness of several colleagues, and if a counter display on the Geiger counter increased by one, and this is documented in before-and-after photographs, etc., then we can be quite certain that the event is not specific to you. We would say that the event must be a part of objective reality, and that this *real* event has become registered in your consciousness, and that of your colleagues.

Even though we can agree that the event was not imaginary, we nevertheless can only be truly certain that *something* happened in the world — that something being, "click!" You created a scenario involving certain materials in a particular region of space and observed over a particular interval of time. And given these constraints, a "yes" presented itself, as in "yes something happened," rather than a "no, nothing happened." After all, at a different place and/or time, the Geiger counter might not have clicked during the interval in which you listened. But in this world, in your lab with the equipment and sample at such-and-such a place and you doing your thing at such-and-such a time, that "yes" occurred — and under your scientifically rigorous interpretation, this one little click blossomed into a narrative that's like a subatomic version of the great American novel.

A World of Stuff

Modern science, particle physics in particular, tells us that the world is ultimately made of stuff — atoms, subatomic particles, particles of light (photons), and so on. This is the standard approach to the nature of things: We are made of stuff, the rest of the world is made of stuff, we can manipulate stuff and observe stuff reacting with other stuff, and isn't all of this stuff interesting? So interesting is it, in fact, that we've tried to figure out how it works for millennia. We've formulated countless theories with physical laws that can predict how stuff will behave, and these theories work incredibly well. It's no wonder that people assume that the world is made of stuff — you can see it, hear it, and feel it with your own senses. And in certain cases where you can't sense something directly, such as the decay of an atom's nucleus, we've built devices that effectively extend our senses: The Geiger counter is able to make something audible that no human could detect naturally. And, according to the "stuff" interpretation of our experiment, that something is an alpha particle that traveled a short distance through the air, entered the detector tube, and initiated an electronic cascade that led to the click coming from the speaker. According to this approach to the nature of things, that individual alpha particle flying through the air was every bit as real and material as your dining-room table.

The "stuff" picture is an intuitive view, for sure: There are rocks and wild animals and air and water in the world, and all of that stuff is made of the chemical elements, and elements are made of atoms and ultimately subatomic particles, and stuff has been around for billions of years, long before the Earth or life. This may seem obvious to you, even self-evident, to the point where you may wonder how there could be any other way to describe the nature of things. However, this intuitive view — "well, of course the world is made of stuff!" — clashes powerfully with certain sophisticated experiments and creates conceptual problems, which I'll discuss in the next chapters.

The picture of a world made of stuff reached its zenith in the very early 20th century, by which time we had learned that even light is composed of particles, which were later called photons. (Albert Einstein won his Nobel Prize for this discovery.) But ironically, this discovery initiated a kind of breakdown in the traditional understanding of a world made of stuff. Over the next few decades, it was found that light and matter can behave as particles sometimes, but under different conditions, they can behave as waves. By 1930, the experimental evidence had shown that no theory could describe stuff as being composed consistently of particles, like tiny billiard balls, that obey the same laws of physics as everyday, so-called classical objects. In particular, physicists discovered that probability becomes important at the smallest scales. (More on these concepts in the next chapter.) This was the dawn of the theory of quantum mechanics, one of the biggest revolutions in the history of science. And, things got stranger: If you can remember your high-school chemistry, you probably know that an atom's electrons form an "electron cloud" surrounding the nucleus. Our best theories say that an electron exists in many places at once, and nowhere in particular — that is, until we attempt to pin down its exact location, at which point the electron presents to us as a specific, stuff-like particle. In the science of stuff, this is how apparently everything behaves at very small scales. So, the "world of particles" picture, as it's described in a modern chemistry textbook or a high-quality science program on television, appears to be not as straightforward as we might naïvely think. Perhaps something much deeper is going on. Is something other than *stuff* at the root of everything we see? Believe it or not, this is a hot question in physics today, nearly a century after quantum mechanics was born.

Consider this audacious quote from the physicist and former White House science adviser John Marburger:

> We can only measure detector clicks. But when we hear the click we say, "there's an electron!" We cannot help but think of the clicks as caused by little localized pieces of stuff that we might as well call particles. This is where the particle language comes from. It does not come from the underlying stuff, but from our psychological predisposition to associate localized phenomena with particles.

Now, someone may read this and ask, if it isn't a particle or other piece of stuff that caused the detector to click, then what did? That's an understandable question, given how we are steeped in the stuff tradition — as well as how routinely, in everyday life, we see things being caused by other things. When we put up a detector and hear a click, naturally we assume that some stuff caused the click. However, it might be like seeing an optical illusion and then asking, "So, why did they make Line A longer than Line B?" The question is ill-posed, as academics like to say. It misses the point, because it inappropriately assumes that our visual judgment of line-lengths is accurate. Two lines may only appear different; we cannot reliably compare the lengths of two lines just by eyeballing them, because sometimes our perceptions deceive us.

Fig. 2. Why is Line A longer than Line B? The question is "ill-posed."

To use another example, if you see a rainbow, it would be inappropriate to ask, "What are the GPS coordinates of the base of this rainbow?" Even though it *appears* that the base corresponds to a specific spot on the ground, there is no such fixed spot (nor a corresponding pot of gold), because the apparent location changes depending on where you're standing. In a similar manner, just because we may conclude that a particle was racing along before it triggered a click, this colorful description has no bearing on whether such a particle *actually did engage* in such a journey, with a beginning, middle, and end.

When a Geiger counter clicks, the only thing we're certain of is: an event happened. Event-information of some kind appeared in the world; we are certain of that. But it's the only thing we really know for sure. Everything else … it gets complicated.

"Just Philosophy"

The last section may have sent up your hackles a little. Maybe you shrugged and said something to the effect of, "This is just philosophy." We've all heard the question about the tree falling in the woods, and some people reduce the millennia-old discipline of philosophy to that moment, or a navel-gazing joke. Particularly when you're young and feel like you have it all figured out, it can seem as if the "real world" is all that matters, and that *philosophy* is mental masturbation or B.S. that never gets anyone anywhere. Even great scientists have given philosophy a hard time; the renowned physicist Richard Feynman quipped, "The philosophy of science is about as useful to physics as ornithology is to birds." The quote got laughs for years — until philosophers pointed out that ornithology would indeed be very useful to birds, were it possible for them to acquire that knowledge.

Funny thing about philosophy: You can't escape it. Almost any statement you can make about the nature of things implies a philosophical framework that defines the assumptions inherent in that statement. This is a subtle point, one easily missed by casual science enthusiasts, but it must be spelled out here.

Consider this: If an alpha particle were indeed making a journey through the air on the way to the Geiger counter tube, it might be impossible ever to know this. We seem to arrive at that conclusion retroactively, once we hear a click, based on a particular philosophical framework and the set of knowledge derived out of it. Philosophers would say that in this framework employed by modern science, particles are a part of the *ontological* picture: Science assumes that these bundles of stuff are present in the world continuously across time, and that they ultimately form the bedrock or foundation of all existence. That's a fine stance for science to take, and it has served us very well overall. However, there is no denying that it is a philosophical stance regardless.

It's a fact: When you think about that particle's existence at some intermediate point of its journey, you're doing philosophy.

Plato's Cave

One of the most enduring concepts of Western thought is Plato's Allegory of the Cave. In this story, prisoners are chained up for life in a cave, their heads immobilized so that the only things they can see are shadows of puppets and other objects on the wall — not themselves or each other. Naturally, since this is all they have ever seen, they consider these shadows to be reality and even give them names. When one of the prisoners gets free and confronts actual reality for the first time, at first he does not believe it, still believing the shadows to be reality. Even though the prisoner is blinded by seeing sunshine for the first time, he needs to go back to the cave and share his discoveries with his old pals — although when he returns, of course, they are skeptical of his claims.

Plato's allegory has been retold many times, but a central theme is: How do we interpret the "shadows on the wall" of the world we see? And if we were shown that our interpretations were misguided, how difficult would it be to accept that truth? For us, the question is: In interpreting reality, and assuming that we live in a world made of objects, have we possibly misled ourselves? What if objects are like the shadows on the wall?

Like it or not, the stuff-based philosophical framework assumes truths where the actual truth cannot be investigated via direct observation or experiment. Any description of the ultimate nature of things can only be arrived at through interpretation, by way of *some* philosophical framework — whether it is particles or bits or what have you. That's just the way it is. "There is no such thing as philosophy-free science," writes Daniel Dennett in *Darwin's Dangerous Idea*. "There is only science whose philosophical baggage is taken on board without examination."

An Alternative Approach: It From Bit

There's another approach to the nature of things that is very different. It proposes that the stuff we can sense around us emerges from something deeper: information. Whereas most of us think of information as being "about stuff," this turns that around. According to an informational account of nature, when we say that an atomic nucleus decayed, the event was actually, *literally* the appearance in the world of information: a "yes" where, under different circumstances perhaps, a yes might not have appeared. There was a probability that the yes would appear, and in fact, in the example presented at the beginning of this chapter, it did appear. According to this second approach to ontology, the description of the particle whizzing through space, the presumptive billions-of-years history of its elementary subatomic components, and all of the other details of the chain of events, is ultimately a matter of interpreting one single bit of information, a click, through the powerful lens of scientific knowledge about the world, which is also information.

Here's an instance where it might seem that I'm pulling something out of my behind (see Introduction). I described the Geiger counter's click as a "yes" and not a "no," like a 1 instead of a 0 on a computer hard drive. I didn't make up that idea; John Wheeler did. A famous 1989 monograph of his, where he advanced a concept he called *it from bit*, is what inspired me to start thinking seriously about information as the basis of reality. "Information, Physics, Quantum: The Search for Links," which is available for free online, is decidedly old-school in its writing style, and its flowery prose and jagged references to other papers can be difficult to wade through. But it is sufficiently rich that I will refer to it again and again.

Wheeler proposed that "every it — every particle, every field of force, even the spacetime continuum itself" derives its very existence "from the apparatus-elicited answers to yes or no questions, binary choices, bits." This is an extraordinary claim. He is saying that the information we have about the world —

for example, whether or not a radioactive sample emitted a particle — does not ultimately derive from stuff. Rather, it's the other way around: Information forms the bedrock of reality, and from that information, a physical reality of *things* emerges. According to Wheeler's it-from-bit conjecture, it is not the case that objects compose the ultimate foundation of the universe. There is only information, measured in bits, at the bottom of it all, and the "its" emerge from the bits, as our minds fill in the gaps to understand some event that we observed to happen:

> The photon that we are going to register tonight from that four billion-year old quasar cannot be said to have had an existence "out there" three billion years ago, or two (when it passed an intervening gravitational lens[1]), or one, or even a day ago ... We ask the yes or no question, "Did the counter register a click during the specified second?" If yes, we often say, "A photon did it." We know perfectly well that the photon existed neither before the emission nor after the detection. However, we also have to recognize that any talk of the photon "existing" during the intermediate period is only a blown-up version of the raw fact, a count.

Coming from one of the world's leading 20th-century physicists, this is a revolutionary statement! Every science book or TV show would describe Wheeler's photon, that little particle or wavelet of light, as flying through space at 386,000 miles per second, traveling on its long journey for billions of years, until finally one day it goes *splat* on a photographic plate or detector on Earth. Wheeler is saying this is not an accurate description: There is a *splat*, but the backstory of an excruciatingly lengthy trip through space is just that. Blown-up backstory.

---

1 A massive object that bends light and provides multiple paths for the photon. The significance of gravitational lensing is discussed in Chapter 2.

Wheeler's paper got physicists thinking: Is nature not fundamentally made of stuff like particles and fields? Is the world ultimately not *objects*, moving and interacting in various ways, but *bits*, instead? It was a radical conjecture. Most physicists today still subscribe to the "stuff" picture, but the issue has become significant enough that in 2013 the Foundational Questions Institute ran an essay contest asking, "It From Bit or Bit From It?" (Winners were roughly divided between both camps.) Digital physics is a growing field.[2]

There are convincing reasons to believe that the world is ultimately informational. To give an introductory example, we are certain that information exists in the world; you are reading some of it right now, and your mind is processing it. As René Descartes put it, *cogito ergo sum* — "I think, therefore I am" — the very act of thinking proves, beyond any shadow of a doubt, that some kind of mind exists to perform that act of thinking. (The exact nature of the mind ... it gets complicated.) Any philosophical framework concerning the nature of things must accommodate the fact that *thinking does evidently occur*, even if we don't yet know exactly how it works. When we think about an apple or particle, despite the arguable ultimate nature of those objects, or the nature of the thinking process, information about them is clearly part of the picture. Therefore, if the "stuff" picture of the world is correct, then the world must contain both stuff *and* information about the stuff. There might

2 Theoretical research into the properties of black holes, particularly by Jacob Bekenstein, has suggested that the contents of a black hole can be described by bits embedded in its event horizon. This insight was also radical, as it was the first time anyone suggested that the contents of a three-dimensional region could be described by information on a two-dimensional surface, or that there was a link between the surface area of a region and how much information can be contained within it (one expects volume to set that limit). Later, an outgrowth of string theory called the holographic principle extended this concept: It suggests that the information contained within the entire volume of the universe is actually "painted" on a two-dimensional boundary, and that our three-dimensional world is projected from this surface, similar to how a 3D hologram projects out of a 2D piece of film. It seems that informational theories of physics are serious, and here to stay.

be an apple, and there might also be our observation of where the apple is, whether it's moving, and so on. This is a kind of philosophical dualism (related to the mind–body problem, discussed in Chapter 9): It says that things have a physical stuff-presence in the world, *as well as* an informational presence. This view extends to the entire universe: There is the star Betelgeuse, and then there is also information about Betelgeuse, encoded into countless photons flung to vast distances, each traveling in a particular direction with a particular wavelength, and all telling something about the stuff within the star that produced them. But this book makes the argument that only the information part is necessary to account for all observations that can be made — no actual "stuff," whatever that might be, is needed. It's a much simpler picture.

I can't discuss information in the natural world without touching on the most iconic example: DNA. The genetic code of living things isn't some kind of abstract information, as you might describe an idea, or a number with a bunch of decimals. DNA is a physical, *recorded* form of information. There are four components of DNA, represented by the letters A, C, G, and T, the arrangement and sequence of which quite literally spells out the genetic code. It's actually a base-four form of digital information, meaning that unlike the binary or base-two data on a computer's hard disk, in DNA, each "bit" can be in any of four distinct states rather than two. Aside from that detail, the genetic code is directly analogous to a string of 0's and 1's that make up a computer file, a digital song, or this book in its electronic form. It's a fact so remarkable that some "intelligent design" advocates have called the genetic code *proof* of an intelligence behind the building blocks of life. While this claim is unfounded (an entire book could be written on that topic alone), the genetic code is undeniably an example of enduring, physical information as a feature of the natural world. Whether it's number-based, letter-based, or chemical-based, it doesn't matter at all; it is still *digital* information, like you'd find on a hard drive.

How Complicated Does the Universe Need to Be?

Now, you're probably so comfortable with the conventional approach to the nature of things that you see no problem with the dualist "stuff + information" view of the world. There are physical photon-objects, and each of these photon-objects carries with it information, such as its wavelength. There is the data represented by the genetic code, and there are also the physical base-pairs in DNA that spell out that code. You've dealt with information about stuff all your life. So what's the problem?

Defining "Information"

From time to time, it will be necessary to define some terms. If you don't define your terms, you can say almost anything you wish; you can use a word one way in one place and another way in the next, in whichever manner most conveniently advances your position. Writing is about clearly expressing what you mean, and using words consistently is crucial to this endeavor.

So, what exactly is information? It's obvious that a table can be made of stuff like protons and electrons, but how can a table be made of information? I have to put off a more complete definition until Chapter 4, but I will elaborate here.

Information is any kind of comparison between two things, such as a difference — or, it can be the *absence* of difference (i.e., the information signifies that the things being compared are equal). Since information can represent any comparison at all, everything that's presumably made of stuff carries with it inherent information. Consider your car: It seems to possess information about its own size and mass, in the sense that we can compare its size or mass to other sizes or masses, and we'd call the results of such comparisons "information." All of the car's components, down to the atoms, seem to possess information as well: An atom of iron has a mass that's a certain multiple of the mass of a hydrogen atom. We don't have to actually measure that number for its mass to be what it is. That information exists in the

This relates to a useful concept in science known as Occam's razor, mentioned in the Introduction. Occam's razor is frequently misunderstood, so let's examine it closely before proceeding. Occam's razor is often stated as, "The simplest explanation is the right one," but that doesn't quite capture the idea, and that definition is easily abused. More rigorously, Occam's razor says that *entities must not be multiplied beyond necessity*. This is a way of saying that if you can get a job done one way with three tools, or another way with two tools, the two-tool method is

world as a detail of basic chemistry, the science of stuff tells us. So, if we were given a sufficient quantity of information, we'd reach a point where nothing else could be said about the car and its components. Under this analysis, we might be tempted to say, "Okay, that's the information *about* the car and its atoms. So, what exactly is the *stuff*, independently of this information?"

I'd try answering, but I have no idea.

It is very hard to wrap one's brain around the idea that a car could ultimately be a collection of information. So, I suggest thinking of the world as a highly advanced computer screen. If a car appears on your computer screen, it's no problem at all understanding that the car is made of information, and nothing but information, in the form of pixels giving off different amounts of red, green, and blue light. But if the screen were good enough — if it had three dimensions, and you could reach in and kick the tires, and your foot bounced off as you expect in "real life" — then you wouldn't be able to tell at all whether the car was ultimately made of stuff or information. (See page 27.) The simplest-case scenario suggests that even the car in your driveway, and all other objects you see in the world, are like that. At the fundamental level, they are made of information and only information, at a sufficiently high resolution that they seem perfectly stuff-like.

better. If you can get it done with one tool, better yet. Occam's razor tells us that a less-complicated scientific explanation involving fewer moving parts is preferred over one with a lot of extra assumptions, conditions, mechanisms, and influencing factors tacked on — and that such a pared-down explanation, often described as more elegant or parsimonious, is more likely to be correct.

The iconic example of Occam's razor being useful in science is the Copernican revolution, when people realized that planets orbit the Sun, not the Earth. To early astronomers, sometimes a planet would appear to turn around and go backward across the sky, which was difficult to understand if all planets orbit the Earth. Astronomers who subscribed to the old Ptolemaic picture of our Solar System had to imagine that not only does every planet follow a generally circular path around the Earth, it also follows paths that are smaller circles on top of circles — "epicycles." Astronomers had to keep adding these epicycles to each planet's orbit in order to more accurately predict its motions and get it to conform to a regular, predictable picture of motion around the Earth. It was assumed that there must be some mechanism driving the planets to move in this complicated fashion. The Copernican system, on the other hand, put the Sun at the center of the motions. This meant that planets had ordinary, simple orbits, with no epicycles or extra mechanisms needed — a much more elegant solution. Today, one could still construct a powerfully predictive Earth-centered model of the Solar System, by having a computer program generate thousands of epicycles for each body. But no scientist would go down that path, because there is no point in describing a situation so complex, when a far simpler scenario is at hand.

Learning to Suspend Belief

I don't expect to convince you that the world isn't made of stuff (not here in Chapter 1, anyway). I merely want you to be open to the informational approach. If you can try to see the world through that lens for the duration of this book, you will be rewarded for doing so. But if you are stuck with the stuff mindset, and if you don't like to read that the human brain isn't always a reliable interpretational tool, then we won't get very far, and there isn't much point in you reading on.

It may sound as if I'm asking you to suspend your disbelief, and pretend for a while that the world is made up of something other than stuff. That's the wrong way to phrase it. Actually, I am asking you to suspend *belief*. Picturing the universe as composed of countless atoms, photons, and whatnot endlessly buzzing about and interacting is a fine philosophical framework — one in which most people believe very strongly. But make no mistake: There is no direct evidence to support it. When we hear a Geiger counter click, we cannot know for certain that a particle was tracing a path through space just prior, or that John Wheeler's photon was racing between galaxies at a time when his nearest relative was a fish. The prevailing Western philosophical framework of long-enduring material particles is a belief system, and for the duration of this book, I hope that you can partially suspend that belief out of a lack of direct evidence, with a curiosity toward seeing the world anew.

Talking About a Revolution

Science enthusiasts who are really on top of things know that physics is in a deep crisis. (Lee Smolin's *The Trouble With Physics*, about how string theory led scientists astray, is an excellent overview.) We don't know why the universe is the way it is, as opposed to some other way; we don't know why time has an arrow that points in a particular direction; and we have no idea how to unify our two most successful theories, quantum mechanics and general relativity. Our mathematical theories are troubled by infinities, despite there being no evidence that infinity is actually a feature of the natural world. The end of John Wheeler's it-from-bit essay cries out for an elegant solution to these tangled riddles:

Surely someday, we can believe, we will grasp the central idea of it all as so simple, so beautiful, so compelling that we will all say to each other, "Oh, how could it have been otherwise! How could we all have been so blind so long!"

These words ring ever truer, as modern physics seeks to add unobservable dimensions and extra universes into its theories, as detailed in Smolin's book. It's going to take a proper revolution of thought to reach Wheeler's goal: a coherent picture of the universe that is not only beautiful, elegant, and simple but also has great explanatory power.

Thomas Kuhn wrote about similar moments in history in his influential book *The Structure of Scientific Revolutions*. For example, before the mixture known as air was separated into its constituent gases, it was assumed that air was an irreducible substance. For centuries, it had even been considered an element of nature. This created all manner of problems when it was discovered that sometimes "air" could make a flame burn hotter (when it was 100% oxygen), while other times it could extinguish a flame altogether (when it was 100% carbon dioxide). A great unexamined assumption will impede scientific progress as effectively as a colossal mass of concrete dams a river.

Even more dramatic than the discovery of gases was the previously mentioned Copernican revolution. Kuhn writes:

Consider ... the men who called Copernicus mad because he proclaimed that the earth moved. They were not either just wrong or quite wrong. Part of what they meant by "earth" was fixed position. Their earth, at least, could not be moved.

It made no sense for Copernicus to claim that the earth moved. By definition, the earth was a stationary thing; this is what people learned as children, and it was plainly obvious and evident to everyone that the earth under their feet was something that never moved, except in earthquakes. Centuries later, even though

Defining "Real," "Objective Reality"

The meanings of "real" and "objective reality" may seem obvious, but they aren't. Consider the widely publicized conjecture that we may be living in an advanced virtual-reality simulation, one that could not be distinguished from the real thing (whatever that might be). Perhaps I am a sentient player in some higher civilization's 3D video game, and if I pick up a rock and drop it so that it cracks open, that event is presented to me as having actually happened. If that were the case, should the rock and its demise be deemed real, or not?

We tend to think of "real" as meaning something like the opposite of imaginary or fake. But this definition is too blunt for our purposes. There presently appears to be no way to tell whether a rock consists of fundamental particles that condensed out of a Big Bang billions of years ago, or is a very accurate simulation of the same. The word "real" becomes useless, even meaningless, in this context. So, if we are going to describe the ultimate nature of *real* things that we can see and touch, we need a more sophisticated definition of reality — a collegiate definition rather than a middle-school definition, if you will.

The term *objective reality* needs only to refer to an internally self-consistent system in which multiple observers agree upon

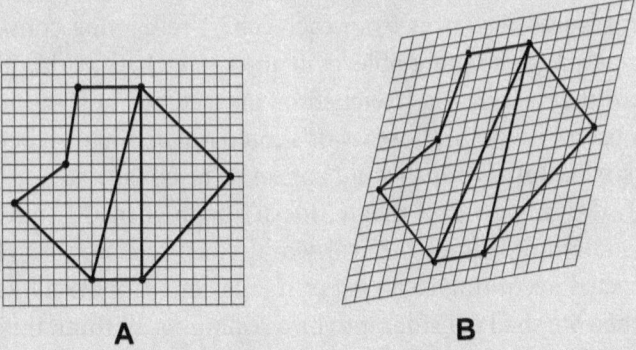

A   **B**

Fig. 3. An inhabitant of **A** would find its world indistinguishable from **B**. The relationships among the points remain consistent, even if the whole graph (or "world") is uniformly deformed. Both situations would seem equally real.

the same course of events. In a world obeying this description, any object, force, or change that is consistently observed shall be called *real*. If you wake up and watch the same news show that millions of others see (as could be confirmed later), and nothing violates the laws of nature, then the events are real, under this definition. We do not need to make any metaphysical statements about the ultimate nature of what is happening, just that it is part of a consistent objective reality. Real events are part of a world where the laws of nature hold, cause and effect don't cross (except maybe on *The Twilight Zone*), and we all agree upon what's physically happening.

We can state a kind of **"reality equivalence principle"**: *A world of objects interacting according to laws is indistinguishable from a sufficiently well-rendered, logically consistent system of information from which objects emerge.* If apples never fly up, and if your keys are always where you left them, then those things shall be deemed elements of objective reality: They are *real*. Whether objects are fundamental aspects of the world, or if instead they emerge from an underlying layer of information, you would not be able to tell the difference, as long as the consistency condition is met. Also, if everything in the world (including rulers) were stretched uniformly, as in **Fig. 3**, objects would still be seen to obey the same physical laws, with their relationships (such as their measured distances from each other) remaining consistent, as compared to the same objects in an unstretched world. In fact, to claim that a world is stretched or unstretched is meaningless, since observers within those worlds could not distinguish between those states. Both worlds would appear to be equally real.

Reality means that magic doesn't happen in this universe. So, if something passes the real-world smell test, and it behaves as expected according to all experiments, and it obeys all known laws, then we shall consider it to be a real thing. Without this definition, labeling something as real or not real is to take an unsupportable stance that can be neither verified nor refuted by experimental test. It is only a matter of opinion.

Defining the word is important: It's the only way we can admit the possibility of there being intuitively *real* things, that also happen to be grounded in something other than *stuff*. Under the more sophisticated definition, some abstract things are considered real: The number line that you studied in grade school is real, because moving two units to the right will always put us at the starting number plus two (i.e., it is logically consistent), and uniformly stretching the whole thing won't change matters at all — it will still be a consistent number line. If you were an inhabitant of the number line, you would have no choice but to consider it a real world, stretched or otherwise.

With this definition, and with the reality equivalence principle, we can move forward with the conjecture that information is the ultimate ground of a world that is in every way objective and real to us.

we learn about Copernicus in school, we don't appreciate the sea-change in thought that was required at the time. Every person who wanted a more accurate understanding of the universe, and their place in it, had to change how they saw the Earth itself.

The takeaway lesson from Kuhn's book is that when the river of scientific progress backs up, a revolution of thought inevitably follows — one that challenges the prevailing way of looking at things. And like a massive dam breaking, rigid monolithic assumptions sometimes get destroyed in the process. So, whether the universe may be ultimately informational, or mathematical (as theorized by Max Tegmark, discussed in Chapter 6), or something else-ical, we need to loosen our assumption that the world is made of stuff, which we assume for really no other reason than because it looks that way.

When you look at an object, you aren't actually looking at the object. You're looking at information about the object. More accurately, you're looking at information from some time ago, due to the finite speed of light. In fact, we can't look directly at an object, nor can any technological device, no matter how sophisticated; *we can only get information.* I'd say that this fact reduces the role of stuff, the object itself, to kind of a ghost, or a shadow. Ironically, in this view, the *object* becomes an abstract thing! The simplest-case scenario would say that this is true of all the stuff in the world. By the end of this book, you'll see how information alone could be the true nature of things, how this view puts all of the pieces together, and how it explains an amazing diversity of mysteries.

Things to Remember From Chapter 1

• Any attempt to describe the ultimate nature of the universe must start with a set of fundamental assumptions.

• In Western science, it's assumed that the world is made of stuff — fundamental particles that aren't based on anything else, just their own existence.

• As experiments grow more sophisticated, it's getting increasingly difficult to square their results with the assumption that the world is made of stuff.

• John Wheeler proposed an alternative, that the world is ultimately made of information, and that things like particles emerge from that information: "it from bit."

• A universe ultimately made of information may be simpler and easier to square with experiments than a universe ultimately made of stuff.

TWO VIEWS OF INFORMATION

e are certain that information exists. By "exists," I mean we know that information is a feature of the world; if it weren't, you could not be reading and contemplating this book, and your parents' genes could not have produced you. What isn't certain is where information comes from — that is the focus of this chapter. It's really a continuation of Chapter 1: Does information derive from *stuff*, the conventional view of information? Does information come from anything at all? When and under what conditions does information appear in the world?

When thinking about these questions, I like to use the analogy of a printed reference encyclopedia. This could be the *Encyclopedia Britannica*, or it could be a book outlining all of the characters and plots of the *Star Trek* franchise, or the *Lord of the Rings* stories, or something completely new. Reference books sit there on the shelf, and if you need to look up something, the information is there waiting to be read. Nowadays, though, there's a new kind of reference: the online wiki. Just as Wikipedia is analogous to a printed encyclopedia, there are countless wiki reference sites on arcane topics, even wikis cataloguing fictional worlds. Unlike one edition of a printed reference book, a wiki can be instantaneously edited and added to; it's constantly changing and being updated. If there is no entry on a topic, it takes only a few clicks to create one. In this manner, active wikis are always growing and becoming more complex.

Given that information is a feature of the world, I'd like to ask: Is information about things more like the information in a printed reference book — meaning it's already out there in some sense, in the form of the properties of the stuff making up the universe — or is it more like the information in a wiki, which is dynamic and constantly evolving in complexity?

Book-World vs. Wiki-World

The conventional "stuff" approach to the nature of things would say that the universe is like a printed book. I call this the "book-world" view. A printed book is made of paper and ink arranged in various ways. In order to get information from the book, you open it up and gaze upon the arrangement of ink on the paper, and in doing so, you pick up information from the pages. The information is already there, of course; it may have been printed long before you were born. But if you want that information to enter your brain, you open the book and read the words, effectively copying the pre-existing information to a new location, your consciousness and memory.

The "stuff" approach says that the universe is fundamentally made of pieces of matter and energy that have been around for a very long time. When we make observations, according to the conventional view, we are getting information from those long-existing chunks of stuff, in the same way you might get information from a book that was printed years ago. If you observe a particle of light from a distant star, you see it coming

Fig. 1. A photon from a distant star (left) is observed to have a specific wavelength and direction. Has this information been a feature of the universe in some sense for millions of Earth-years? One view says yes, another says no.

from a particular direction and can measure its wavelength (color), which is kind of like making a copy of the information that had been carried by the photon, and calling it your own. Naturally, the conventional approach to the nature of things assumes that since everything is made of stuff, that photon of light must have had those properties all along in some sense — direction/wavelength information was encoded into the photon at its distant source, the star. And, that information would be a feature of the universe whether we intercepted and measured the photon or not, just as the information in any printed book would be there whether or not you ever opened it to read it.

We can make a little picture of this view of stuff and information: There is us (made of stuff) and there is an object (also made of stuff), and in observing that object, information inherent in the object is transferred or copied from the object to us:[1]

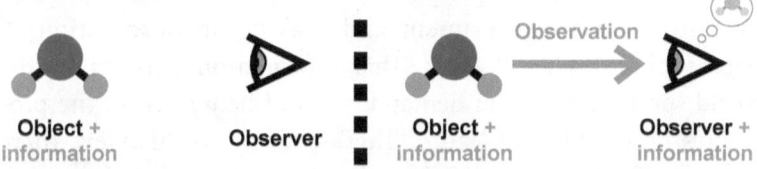

| Object + | Observer | Object + | Observer + |
| information | | information | information |

Fig. 2. In the book-world view, objects carry information with them (left). An act of observation involves creating a copy of that information (right).

Now let's consider another view of information, which I call "wiki-world." Since you have been steeped in the book-world picture of information your whole life, it might be a little harder to imagine what wiki-world information might be like.

Imagine a wiki that's similar to the reference wikis online, but not exactly. This wiki is about a completely new fictional world. Let's call it *Reign of the Mantelopes*. There are plotlines and characters just as in *Star Trek* or *Game of Thrones*, but there are no books or movies about this fictional world; the wiki itself

---

1 I am not claiming there is anything special about the mind or about human observation. If you have a machine that can measure some property of an object and then print that information onto a piece of paper, the principle is the same.

is the authority on the topic. Second, this wiki is not written and revised by human editors. Instead, it pretty much "writes itself," with all of its details developing over time based on the things that people look up. Suppose you're a fan and you want the answer to a question: "I know so-and-so is a mantelope, being half-human and half-antelope. So, is his mother human? Yes or no?" With a human-written wiki, either someone would have entered that information, or not, so you'll either get the answer or you won't. This wiki we're imagining, though, will give you an answer regardless. If no one had ever asked that question before, the wiki automatically chooses an answer at random. And from that point on, this information is a detail about the character and the *Mantelopes* world in general.

Experiments in physics are consistent with the proposition that the universe is similar to a wiki-world like this. According to this view, when we "query nature" — that is, we ask a question by setting up an experiment and making an observation — nature provides an answer. This information appears in the world spontaneously, on demand, without being tied to the properties of any kind of "stuff." That's because in this view, there is no stuff! This is what John Wheeler meant when he wrote: "What we call reality arises in the last analysis from the posing of yes–no questions ... all things physical are information-theoretic in origin and this is a participatory universe."

It's revolutionary for Wheeler to claim that our universe is participatory. He's saying that we play an active role in bringing about the information in the world. He's also saying the universe doesn't just contain information; all of the things that we call physical *are* information. Rather than merely copying what's already there, we actively participate in bringing about the information, through making observations (querying nature). This might seem like only a fanciful idea, and perhaps easily dismissed, but it's supported by well-known experiments. Lots of them.

The next few sections deal with these important experiments. I've tried to explain them as clearly and accurately as possible, without oversimplifying or altering the key facts. Read through the descriptions of the experiments while keeping in mind the alternative view that information in the world does not necessarily pre-exist, like the information in a printed encyclopedia. Later, I'll come back to the wiki-world picture and put it together with the experiments. In the mean time, you might begin to wonder why you ever believed in stuff in the first place.

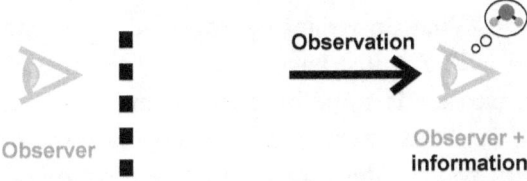

Fig. 3. In the wiki-world view, observation causes new information to enter the world. For observers, objects appear to emerge from this information.

The Two Faces of Stuff: Particles & Waves

Physics has kind of an obsession with particles. There's an entire discipline known as particle physics; its researchers have developed the Standard Model, which details the various fundamental particles and their properties. Even gravity and cosmic inflation have been associated with particles, called the graviton and inflaton. When John Wheeler mentions a photon arriving from a distant galaxy, naturally we imagine a little packet of stuff racing through space. But even within the conventional view of a world made of stuff, particles are only half the story. That's because every particle has an alter-ego twin, a corresponding wave. Sometimes the stuff that we call an electron or a photon behaves like a particle, and sometimes it behaves like a wave. Physicists refer to this dual quality of matter and energy as *wave–particle duality*. Bizarrely, if you set up an experiment to look for waves, you'll observe waves; if you set it up instead to observe particles, you'll observe particles. This explains why Isaac Newton thought that light consisted exclusively of particles or "corpuscles," but a century later, Thomas Young demonstrated that light is made of waves, and a century later still, Albert Einstein showed again that light is made of particles (he won a Nobel Prize for that discovery). A couple of decades after that, the emerging field of quantum mechanics had to concede that Newton, Young, and Einstein were all correct: Photons of light, electrons, protons, and all other bits of matter and energy can behave as either waves *or* particles. Which form you observe depends upon the way that you set up the experiment.

What do we mean when we say "behave" as a particle or a wave? You just have to think of the meanings of those words. A particle is something that's small and localized, like a grain of sand but much smaller. A wave, on the other hand, is spread out. You can describe the precise location of a grain of sand, but you can't do the same for an ocean wave. You could say that it's 100 feet from shore, but you can't point to one specific seashell that will get wet when the wave breaks; it's spread out, maybe for hundreds of yards, and it wets a lot of things.

Let's say you want to test whether a photon is more like a particle or a wave. You can do this by giving the photon a chance to take one of two possible paths, or maybe both. A particle, since it's a localized thing, can take only one path or the other, in the same manner that a grain of sand will either get blown over a fence, or not, but never both. Waves, being spread out, don't behave this way. If an ocean wave strikes the end of a wall that's perpendicular to the beach, the wave will split in two,

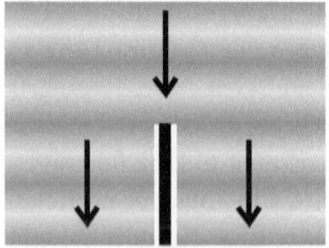 

Fig. 4. A wave (left), being spread out across space, can take both paths around a barrier, but a localized particle (right) can take only one of the two paths.

and it will take both paths. For photons in particular, the iconic way to present two possible paths is with a "double slit," which is a plate with two very thin slots placed close together. Fire photons at a double slit, and if they're particles, each particle will pass through one slit or the other. If they're waves, each spread-out wave will go through both slits. Another way to give photons two possible paths is to use something called a half-silvered mirror, which lets half of the light through and reflects the rest. Since a photon represents the smallest possible chunk of light at a given energy, a particle-behaving photon will either pass through or be reflected, but not both, because it can't split in half. A wave-behaving photon, on the other hand, will be able to take take both paths.

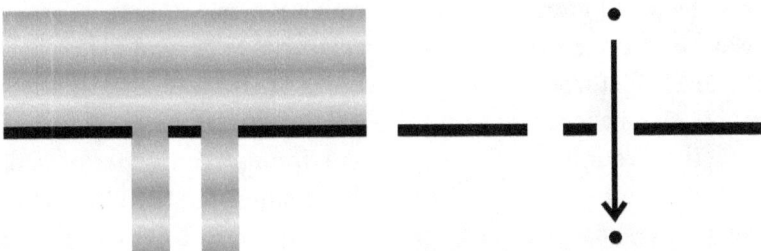

Fig. 5. As in Fig. 4, a wave (left) can take both paths through a double slit, but a particle (right) can pass through only one of the two slits.

How can you tell whether photons take only one path or both paths? You place a screen some distance away from the double slit and see if anything unusual happens. (You can do this experiment yourself at home with a laser pointer. Look online for instructions.) Rather than getting two spots on the screen, which is what you might expect from sending light through two slits, you'll see many stripes or spots instead. This suggests that the light arrives at the slits not as particles but as waves, which strike both slits at once; then each slit produces a separate wavelet, and as the two wavelets combine again at the screen, they interfere with each other. On some places on the screen, wavelets arrive

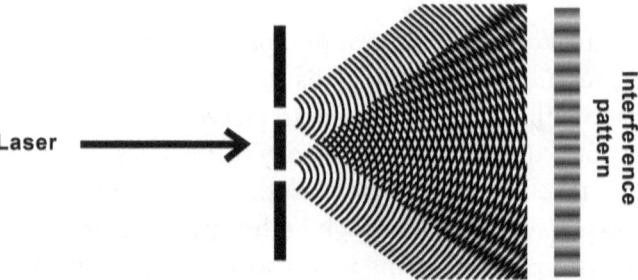

Fig. 6. When light waves go through a double slit, they interfere at the screen (far right), producing bands of light and dark. (Not drawn to scale.)

that reinforce each other, while nearby, the wavelets cancel each other out. This produces what's called an interference pattern, with alternating areas of light (where the wavelets reinforce) and dark (where they cancel).[2] You can observe a similar phenomenon by passing waves of water through two gaps: Some distance away, the wavelets pile on top of each other in certain places and the level fluctuates, whereas maybe an inch away, the wavelets cancel each other and the surface is still.

This is similar to the experiment Thomas Young performed. To complete the demonstration, Young simulated the effect with water waves, as described above. His experiment convinced the scientific world that Newton was wrong and that light consists of waves, not particles. After all, if light were particles, then each slit should allow only a narrow beam through, like arrows making their way through slots in the wall of a castle. The particles should strike the screen in just two spots, or perhaps smeared out evenly. But that's not what happens. The light produces a predictable pattern of interference exactly as if it had passed through the slits as waves.

The year was 1803. Photography hadn't been invented yet — but if it had, it could have turned Young's discovery on its head.

---

2 With one slit open, a more subtle wave-like pattern still appears, due to an effect called diffraction. If you hold two fingers very close together in front of your eye, and then focus through this single slit at something bright, the gap will blur out into a series of light and dark stripes. That's diffraction in action, and you just demonstrated the wave nature of light. You actually don't need a double slit to do that.

That's because if you replace the screen with a sensitive photo-graphic plate, and then you dim the light to an extreme degree so that it's passing through the slits one photon at a time, the light will appear on the plate as particles! Each photon will produce a single tiny dot, which is just what you'd expect if the plate had been struck by a particle. All of this happens with electrons and other minuscule beasties, too.

It gets weirder. Take the one-photon-at-a-time experiment above, but let it run so that lots of photons build up on the plate. Can you guess what happens? At first, the particles seem to appear randomly. Then it becomes evident that particle-spots are piling up in some regions more than others. Finally, a pattern of stripes emerges — exactly the same pattern of interference that you see when you blast trillions of photons through a double slit at home with a laser pointer.

These results are counterintuitive. If a photon hitting a pho-tographic plate produces a particle-like spot, naturally you'd assume that the same photon went through the slits as a particle, too. But, the interference fringes are evidence that *something* en-countered the slits as a wave. To explain how particle-like photons contribute to the buildup of an interference pattern, sometimes it's said that each photon-as-particle "interferes with itself" or "goes through both slits at once." Perhaps more commonly, though, the photon is said to go through the slits as a wave, but it ends up on the plate as a particle. A photon can be seen behaving like a wave or like a particle, but never both simultaneously.

Fig. 7. Even when particles are allowed to build up on a photographic plate gradually, they form a wave-like interference pattern. For want of a better explanation, it's sometimes said that each particle "interferes with itself."

Welcome to the world of quantum phenomena! People often say that quantum behavior is weird or even impossible to understand, but this might only be because the stuff approach to the nature of things has met its match. If you let go of that persistent idea of stuff, the weirdness melts away, as we'll see.

Countless variations on the double-slit experiment have been done over the decades, and they always produce consistent results. It can seem as if the more sophisticated the variation, the further down the quantum-weirdness rabbit hole we go. For starters, consider the following thought experiment: Suppose we'd like to catch photons in the act of going through either one slit or the other. So, we aim a device at one of the slits, or both of them, to "see" the photons passing by. (This is a thought experiment because it's impossible to detect a photon without destroying it in the process.) Quantum theory predicts that the pattern of stripes on the screen should disappear in this case, because otherwise the photons would be simultaneously behaving as particles (by being observed taking only one of the two possible paths) and as waves (by contributing to an interference pattern). In fact, in recent years, physicists have devised clever ways to detect particles indirectly or measure them extremely weakly — and the results confirm what the thought experiment predicts. It seems that if anything is determined about the photons in transit at the slits, the wave-like interference pattern on the screen has to diminish. Even physicists who have never stepped outside of the stuff tradition tend to put this bizarre effect in informational terms: If we have no information about which path the photons took, then it will appear that they traveled through the slits as waves. However, if we seek the "which path" information, then each photon indeed appears to go through only one of the slits, and no interference pattern forms on the screen. Furthermore, if we get only *partial* information about a group of photons, we get a partial pattern. Merely the act of seeking information, or potentially obtaining information, alters the way that the stuff of the world behaves.

Whether things seem to behave as waves or particles depends on how much information we extract from them!

Well, isn't that a kick in the pants. And truly weird, if you believe that the stuff of the world behaves differently depending on whether we or our devices are looking at it.

Enter Probability

In the case of individual photons slowly building up an interference pattern, one has to wonder: What on earth or in the universe is steering each photon to a position on the screen which, when analyzed in the context of a lot of other photon impacts, suggests that each photon once lived as a wave?

The standard explanation is that when you set up a double-slit experiment with the slits a certain distance apart and the screen is this far away and you use light of a particular wavelength, you create a situation where the final resting place of any photon depends upon a corresponding probability distribution at the screen. In one location, there's a high chance that a photon will appear and produce a spot; a short distance away, the probability is low. This is similar to ordinary probability distributions that you're familiar with: Given a large group of people, a few will be very tall or very short, but most will be somewhere in the middle. The probability distribution can be represented with a "bell curve," with the average height being somewhere near the curve's peak. It's similar for the photons on the screen, except that the interference effect produces lots of peaks and valleys:

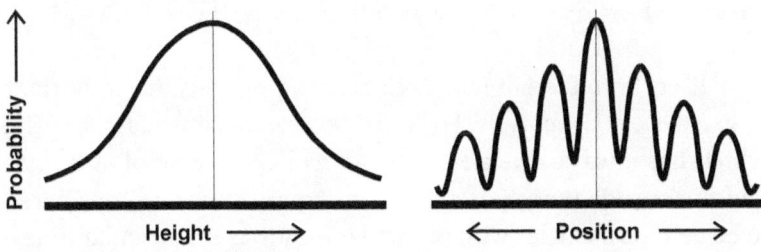

Fig. 8. The probability of any person being a certain height (left) is similar to the probability distribution at the screen in a double-slit experiment (right).

All of this business about probability may sound familiar, if you recognize the term "electron cloud." At first as students, we learn that the electrons in an atom orbit the nucleus. Then in high school, we learn something bizarre: Electrons don't actually, physically go around the atomic nucleus like planets around the Sun. In fact, such an act would be impossible. Electrons have negative electric charge, and anytime a charged object moves, it creates electromagnetic waves, which are energy. An electron physically orbiting a nucleus would therefore radiate away its own momentum, so all atoms would collapse almost immediately and there wouldn't be any atoms. Instead, chemists and physicists describe electrons as forming an electron cloud consisting of *orbitals* around the nucleus. Just like the bell curve of height, or the probability distribution in a double-slit experiment, an orbital is a probability distribution, and that distribution tells us the probability that we would find an electron at a particular location, if we looked there.

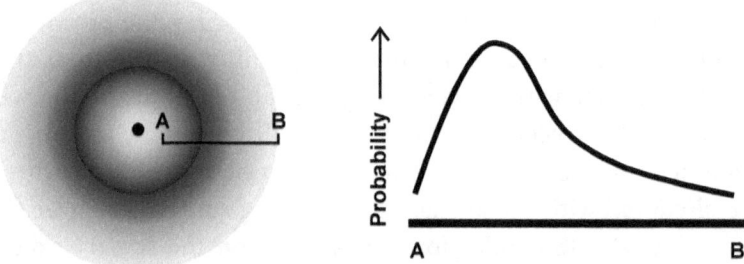

Fig. 9. In an electron orbital (left), there is a probability of finding an electron somewhere between A and B. This probability is shown in the graph (right).

Even if you're familiar with electron orbitals, it's important to grasp the concept fully. For some 90 years, mainstream science, which has always assumed that the world is made of stuff, has had to concede that the electrons in atoms cannot be considered to be defined particles with particular locations at particular times. Instead, they blur out into a probability distribution. Only when you try to pin down an electron's location does it show up as a defined particle, the same way a photon in a double-slit experiment manifests as a localized particle only when you inquire which slit it went through, or when it goes *splat* against the screen. As it turns out, quantum physics tells us that *any* "particle" that we haven't observed — but which is there, at least in some sense — can only be considered a wave of probability. (Physicists call this a *wave function*.) You're more likely to find a defined particle when you look in some places, but there is always a small chance that it will appear nowhere near where you expected it to be.

A scientific theory always gets a boost when it's exploited by a practical application, and the particles-as-probabilities idea is a good example. An electronic component known as a *tunnel junction* takes advantage of it. In a tunnel junction, electrons travel near an insulating barrier that they cannot pass through in the classical particle sense. Regardless, each electron's probability distribution *does* reach across the barrier, and as a result, occasionally an electron can be found on the other side. In this phenomenon, known as *quantum tunneling*, the electron is said to "tunnel" through the otherwise-impermeable barrier, but actually, "appearing on the other side" may be a better way to describe it. The likelihood of observing an electron cross such a barrier can be predicted very closely via the electron's probability distribution or wave function.

Stuff as a probability distribution, rather than particles? The world of stuff just gets curiouser and curiouser.

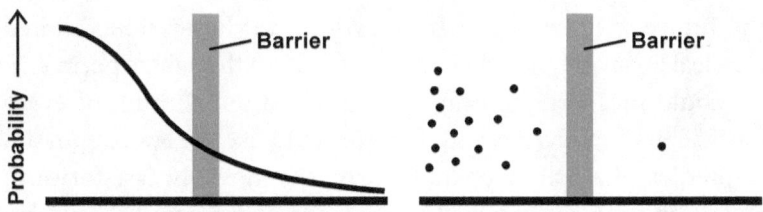

Fig. 10. In a tunnel junction, electrons that could not ordinarily cross through a barrier can be found on the other side regardless, simply because there is a small probability of finding them there.

Taking Snapshots With the Uncertainty Principle

Suppose you lose a ball in the backyard, and you go looking for it, and you find it. The ball will be in one precise location, and it'll probably not be moving, at least in relation to your backyard. This is how the world of ordinary or *classical* objects works: The position and speed of any classical object can be measured or described to any degree of precision we like. There's nothing unusual about that; it's one of the things we mean when we say an object carries information. (The ball is 12 feet southeast of the base of the orange tree; its speed relative to the tree is zero.) This is not true, however, of very small objects such as photons and electrons. It turns out, there is a limit on how much information we can get from a quantum or non-classical object. If we choose, we can measure an electron's position to a very fine degree of precision. We can also measure its speed or momentum to any degree of precision. However, we can't do both at once, in the same way that we can't observe wave behavior and particle behavior at once. If we tightly measure an electron's position, then its momentum can only be somewhere within a probability range; if we tightly measure its momentum, then its position becomes a blur. Unlike the ball in the backyard, at very small scales, nature prevents us from obtaining complete information about an object. This is called the *uncertainty principle*, discovered by Werner Heisenberg, one of the founders of quantum theory.

The uncertainty principle is a big deal for a lot of reasons. For one thing, it's a way that nature prevents us from predicting the future with certainty. If everything in the world exhibited classical behavior, then theoretically (i.e., in a thought experiment) we could measure the exact position and momentum of every particle in the universe, and then we could use the laws of physics to predict what will happen. Just run the movie in fast-forward, and at any future time we'd know where every atom should be, and therefore, we'd know the future exactly. Turns out, we can't know everything about every particle. But if the world is made of stuff, it's not entirely clear why this should be the case.

Robert Lanza has a good analogy for describing the uncertainty principle. Imagine an archery contest, with a photographer on the scene. An archer draws the bow and shoots an arrow toward the target. As the arrow flies past the photographer, he snaps the shutter. What will the photo look like? Well, that depends upon how the photographer sets his camera. If he uses a very fast shutter speed, the arrow will appear very sharp, and by looking at the photo, you can tell exactly where in space the arrow was when the shutter was triggered. However, there's a downside: You can't tell anything about how fast the arrow was going. So, the photographer slows down his shutter for the next shot. Now the arrow is a blur — which is useful for calculating its speed (just divide the length of the blur by the time the shutter was open), but now there is less information available about the arrow's position. This analogy isn't exact; an arrow is a classical thing with a definite position and speed at all times. In the uncertainty principle, it's as if the arrow were only the *possibility* of an arrow — and when someone clicks their camera, the arrow shows up in some form, depending on what information the person is seeking.

The uncertainty principle seems to suggest that making observations of quantum objects is like taking snapshots of reality, those snapshots showing up as particles. We can change the specifications of the snapshots depending on how we choose to observe. Before the snapshot, an electron is a probability distribution; afterward, it's a particle, and the snapshot provides information about that particle. Before the snapshot, the electron exhibits wave-like qualities; afterward, the wave seems to collapse into a definite, localized thing whose position or momentum we can determine precisely (but not both).

We don't have to consciously intervene to find particles, and humans are not the only things that can make observations. After all, each particle-speck that appears on a photographic plate can be considered an observation of a photon by a minute photosensitive silver-iodide crystal. Particles also leave trails in mica crystals, which can be seen under a microscope; a physicist would say that in the course of this interaction, the mica "observed" the particle. Regardless, some kind of observation or direct interaction — whether animal, technological, chemical, or mineral — seems to be the only means by which particles make any appearance whatsoever in the world. In the absence of observation-like interactions, the things of the universe go about their shadowy, probability-based business, apparently in a very non-particle-like manner.

If things are probability-based before they are observed in some way, how do we have license to say that these things consist of *stuff*? The evidence requires us to re-think what we believe this stuff is all about. Also, when we go looking for information about a particle's position or momentum, and then we get that information, how can we say that the information pre-existed in the thing before the observation, the way words and sentences pre-exist in a book before we open it for the first time? The thing was only a probability — that's what the experiments demonstrate again and again. It's a little like asking where the information is about the outcome of a roll of dice that has not yet occurred. If there were no uncertainty principle, and therefore every particle could be assigned an exact location and momentum, we could say that the information about the future dice-roll is encoded in the positions and momenta of all of the particles making up the dice, the table, the player's hands, etc., and in principle we could predict the outcome of the roll using that information. The uncertainty principle ensures that these kinds of predictions are impossible. Until the dice are actually rolled, all of that information, and hence the information about the roll-to-be, is uncertain.

Can we say that information is a feature of the world, in the case when it's represented only by a probability? That's what quantum physics has made us confront. Let's look at a few more experiments, because here's where it really starts getting interesting.

Delayed-Choice: Messing With Causality

The delayed-choice experiment, first suggested by John Wheeler, is a variation of the double-slit experiment. Wheeler wanted to know whether there is a point in time when a photon (or electron, etc.) "decides" to behave as a wave or a particle. Perhaps if a photon encounters two slits, it will travel through as a wave — which will manifest in the interference pattern seen when a lot of similar photon impacts are allowed to build up. Conversely, if there's only one slit, or one of the slits is being monitored for particles, then there is only one possible path for any individual photon, so maybe in that case each photon travels through as a particle.

Wheeler asked, what would happen if you changed the experimental setup *after* a photon made its way through the slits? Initially this was Wheeler's thought experiment, but in recent years, experimental variations on this idea have actually been performed — and the results challenge our assumptions about time, the past and future, and cause and effect.

Consider the half-silvered mirror mentioned earlier. As a particle, a photon can take one of two possible paths (reflected or passed through). As a wave, the photon can split into two wavelets and therefore take both paths, so when we recombine the paths and send lots of photons through the apparatus, they should accumulate to form a striped interference pattern. However, if

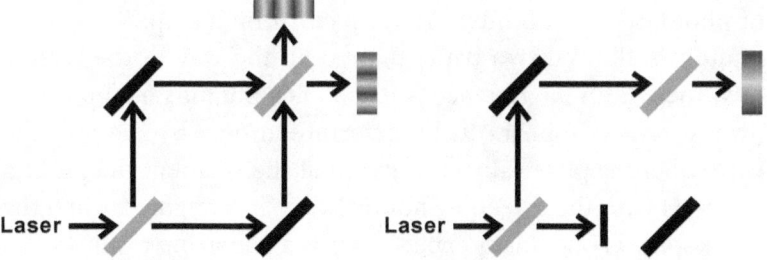

Fig. 11. Splitting and then re-combining light waves with half-silvered mirrors (gray) produces interference, as in the double-slit experiment. Blocking one of the paths after the first mirror (right) destroys the interference.

you do anything to determine which of the two paths the photon took, the interference pattern will go away. For example, if you block one of the paths after the mirror, then there is only one path that each photon can take to the screen — and the apparatus seems to know this, because firing photons one at a time will cause a buildup of specks with no interference stripes. (The same thing happens if you cover one of the slits in a double-slit experiment.) Since this effect occurs with individual photons, the probability distribution for any individual photon impact on the screen must be different when one of the paths is blocked. Without the obstruction, though, the specks build up as interference stripes, each photon leaving evidence that it navigated the mirror as a wave. Once again, this shows that if you set up an experiment to look for waves, you'll find evidence of waves, and if you set it up to look for particles, you'll find evidence of particles.

So, what would happen if you fired a photon at the half-silvered mirror, and *then* you decide how many paths it can take? For example, the paths can be long fiber-optic cables, one of which has an electronic shutter on the end. Using this method, physicists can change the experimental configuration after the photon has already encountered the mirror, the location where it seems to decide whether to behave as a wave or particle. The results of actual experiments confirm Wheeler's prediction: If a photon is given access to two paths, but then we close one path after it passes the mirror, and we repeat this process for a lot of photons, then no interference pattern builds up. But if each photon is allowed two paths that go all the way to the screen, then there *is* an interference pattern. Each photon's final position on the screen appears to be determined *only* by the way the experiment is physically configured at the moment the photon interacts with the screen — not by how it's configured when the paths split, as one might expect. This is an amazing result!

To take things to the extreme, John Wheeler imagined a cosmic-scale version of this experiment. Massive structures such as galaxies cause light passing by to bend. If a galaxy is situated between us and a far more distant light source, like a quasar, a photon-as-particle could follow a path that bends around one side of the galaxy, or it could bend around the other side. Alternatively, a photon-as-wave could take both paths. In this respect, the galaxy serves the same path-splitting role as a double slit or half-silvered mirror. Wheeler surmised that if we combined the paths, we would find interference (wave evidence), and if we didn't, we would find no evidence of waves. Even at this astronomical scale — and even though the photon may have passed the intervening galaxy before Earth was even a planet! — the photon's behavior ultimately seems to depend on whether the lab equipment on Earth is configured this way or that way. Wheeler's thought experiment seems to imply *retrocausality*, where a human being's decision today can alter the behavior of stuff in the universe going back billions of years.

I don't have to tell you that's crazy talk. But even though the cosmic version of the delayed-choice experiment is presently unfeasible, the other versions all suggest that Wheeler's predicted outcome would, in fact, be correct.

Quantum Eraser: Making Information Disappear

Quantum-eraser experiments are other variations on these experiments. Here's the basic idea: Photons (or electrons, etc.) are given the opportunity to take one of two paths, or both, and then anything that takes one of the paths is "marked," to identify

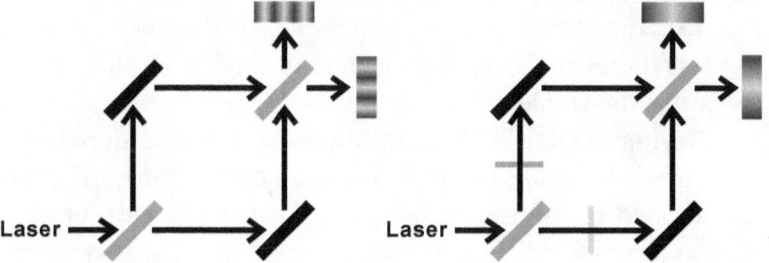

Fig. 12. "Marking" the photons with polarizer filters (thin gray lines, right) destroys the interference patterns. These examples are simplified slightly.

that it took that path. For example, photons passing through or reflected by a half-silvered mirror can be sent through two different polarizing filters. These will orient the wavelets of light differently between each path, so that by recombining the paths and measuring the polarization of any individual photon, you could determine which path it took. Not too surprisingly, marking the photons in this way eliminates any interference in the combined light beam, or in the pattern of photons built up one at a time. This occurs even when the marking is done cleverly and in a way that does not directly impact the photons at all.[3] It is normally said that we have extracted "which path" information on the photons, and this is what physically changes the situation. But actually, we don't even need to collect the information in order for the photons to exhibit the particle behavior of taking only one path: The information only needs to be *potentially* available, in the sense that that some physical aspect of the experimental setup makes it possible to distinguish between the paths taken. If the path-takers can be distinguished from each other even in principle, then they won't interfere. Individual photons don't "interfere with themselves" in this case; each photon appears to traverse only one side of the apparatus, behaving as a particle, not a wave.

What's more interesting is that if the markings are subsequently *removed* — if the "which path" information is destroyed or made physically unavailable at some point before the photons are recombined — then the interference pattern *returns* (see **Fig. 13**). The photons behave as if they had never been marked at all, as if they had been waves going through both sides of the apparatus the whole time.

Trying to analyze these findings using conventional wisdom, it would *seem* as though at some point of time, information existed in the world that pinned down the path of each photon. From the other experiments, we've seen that extracting

---

3 The most definitive quantum-eraser experiments leave photons untouched by marking their entangled partners instead. Entanglement is discussed later in the chapter.

"which path" information, or making "which path" information appear in the world, has seemingly caused wave-photons to collapse into particle-photons. In the case of delayed-choice experiments, making this information appear in the world seems to retroactively cause the photon to act like a particle all along. But now we're erasing this information, and in doing so, we're wiping the slate clean again: Now it seems that we're retroactively causing the photon to act like a *wave* all along. History has been revised, and revised again.

For those steeped in the tradition of stuff, it's no wonder quantum physics is often described as impossible to understand. The quantum eraser can make you want to rip the hair out of your head: *Wait a minute, we marked the photons with the polarizing filters. Why should subsequently erasing that information change the way each photon behaved from the beginning?*

I will argue that it's because we don't take information seriously enough. We're still intent on thinking of information as coming from some underlying stuff. We insist on thinking that information just reflects real objects, even though these real things seem to change their properties — even retroactively! — depending upon the whims of how we choose to experiment on them. We speak of the "behavior" of quanta such as photons because we assume that even in an unobserved system, there must be objects there, and they must be behaving in some specific manner or another. We might want to let that assumption go.

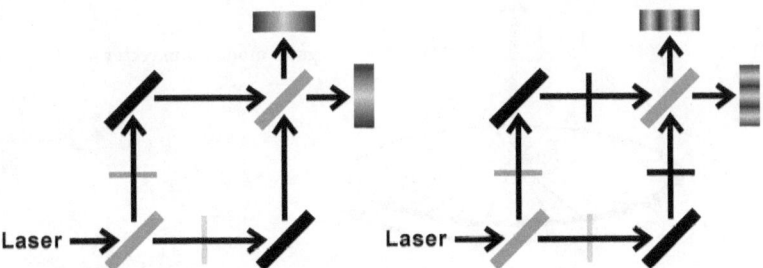

Fig. 13. Erasing the markings by adding identical polarizer filters to both paths (black lines, right) restores the interference patterns.

Spinning But Not Spinning

In physics, momentum is related to the amount of energy carried by an object's motion. A baseball flying through the air has momentum, which we know when the ball hits a window: The energy breaks the glass. Momentum doesn't like to be changed, either in magnitude or direction, which is why an object in motion tends to stay in motion, and an object at rest tends to stay at rest, unless acted on by other forces.

Angular momentum is a measure of the rotational momentum of an object that's spinning. Like linear momentum, angular momentum doesn't like to be changed, either; a spinning object wants to stay spinning, and it wants to stay spinning along the same axis. This is why a bicycle is easier to balance when the wheels are turning, and it's what keeps a spinning top from tipping over. Physics uses a handy mathematical convention to describe how much angular momentum an object has. If a wheel is rotating, its angular momentum can be represented as a line (called a *vector*) coming straight out of the axle. The vector's orientation shows the orientation of the spin (the line coincides with the object's axis of rotation), and its length represents the magnitude of the momentum. If you slow down the rate at which the wheel rotates, the vector is shorter; speed it up, and the vector is longer. The vector is also proportional to the object's mass: A wheel that's twice as heavy but spinning at the same rate would have an angular-momentum vector that's twice as long.

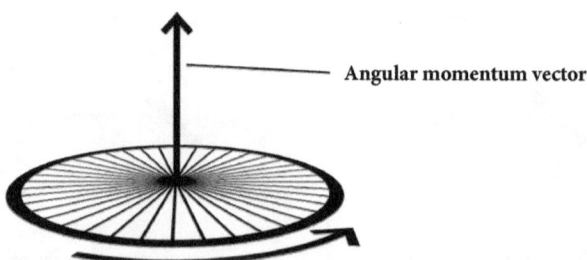

Angular momentum vector

Fig. 14. The angular momentum of a rotating wheel is represented by a vector along the axis of rotation. A longer vector means more momentum.

That is so-called classical angular momentum, which involves everyday objects. The angular momentum of a spinning bicycle wheel or a planet can be pretty much anything; the vector can be any length, and it can point in any direction. Very small things like electrons can also have angular momentum, but it's different. This kind of angular momentum can take only very specific values. You can't make an electron, for example, spin a little faster or a little slower, as you can with a classical object. We say that the spin is *quantized*; its angular-momentum vector can be only a very specific length, and nature doesn't allow it to have any other length. For this and other reasons, physicists don't believe that electrons and other particles are actually, *physically* spinning on an axis, like a wheel or a planet. Since particles such as electrons always spin at the same "rate," and this property of the particle never changes, the property is sometimes called *intrinsic spin* or *intrinsic angular momentum*. Still, for the purposes of visualizing what's going on with these particles, it can be helpful to imagine them as physically spinning on an axis.

One of the most important experiments of the last century, which proved the existence of intrinsic spin, is called the Stern–Gerlach experiment. An atom with one electron in its outermost shell, like an atom of a metallic element, has an overall intrinsic spin that causes the atom to respond to a magnetic field. Suppose we prepare a stream of silver atoms and fire them at a screen. They pass through an electromagnet, but for now it's switched off. The atoms will travel in a straight line and strike a small spot on the screen. Now, if we start to apply power to the electromagnet, creating a magnetic field that the atoms are passing through, something interesting happens: We'll start to see the spot on the screen split in two. If we make the magnetic field stronger, we'll see the stream of atoms separate into two sub-streams, creating two distinct spots on the screen. The atoms are being sorted by the direction in which their outer electrons are "spinning." For some, the angular-momentum vector points in one direction — these atoms veer one way in the field, producing one of the

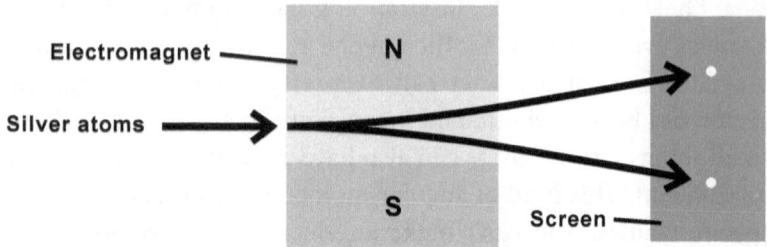

Fig. 15. In the Stern–Gerlach experiment, atoms are sorted by their outer electrons' intrinsic angular momentum. The fact that two distinct spots appear at the screen proves that this electron spin is quantized.

two spots. The rest of the atoms veer the other way. All of the electrons veer the same distance one way or the other, which demonstrates that they are all "spinning" at the same exact "rate," just in precisely opposite directions. If the magnetic field is oriented in the up–down direction, we say that half of the atoms are *spin-up* and half are *spin-down*.

So far the situation doesn't seem that unusual. You can pour a bucket of pennies on the floor and proceed to sort them, finding that half of the pennies are heads-up and half are heads-down. However, as with the quantum experiments we've already examined, counterintuitive things happen when we vary the experimental setup.

If you take the same setup but you rotate the electromagnet by 90 degrees, now the apparatus will sort the atoms into spin-left and spin-right versions. If you point the apparatus toward the ceiling, you will separate the stream into spin-this-way and spin-that-way atoms. No matter how you choose to orient the equipment, you will always split the atoms into two streams that deflect by the same but opposite amount. Why do we always observe the atoms at the screen as two distinct spots? If we were to attempt this trick with classical objects, say, by throwing randomly oriented refrigerator magnets through an electromagnet, the initial orientation of each individual magnet would affect its position at the end, and with a large number of magnets randomly oriented, this would produce a random distribution. Some of the magnets would deflect by a lot (if their initial orientation lines up with the electromagnetic field), some by only a little, and some none at all (if their initial orientation is perpendicular to the field). This would produce a smooth, uniform distribution on the screen, and that's not what we see here: We see two tight spots. Stern–Gerlach was the first experiment to show definitively quantized, non-classical, bizarrely either–or behavior in small objects such as atoms.

Now suppose we follow one Stern–Gerlach apparatus with another one, oriented differently. First we split the stream into spin-up and spin-down atoms; we then pass, say, only the spin-up atoms through another apparatus that's oriented perpendicular to the first. Would it separate them into spin-left and spin-right versions? Yes, that's exactly what happens: An atom that's identified as spin-up can then be re-identified as spin-left, or spin-right; there's a 50% probability of getting either result. That's weird enough — how can a group of atoms all "spinning" in one direction get split according to their spin measured in a perpendicular direction? Stranger still, if we add a *third* apparatus at the same orientation as the first, we find that the spin-left atoms (for example), which previously had been all spin-up atoms, can now be split 50-50 into spin-up *and spin-down* atoms (see **Fig. 16**). Imagine that — an individual atom can be measured as being spin-up, then spin-left, and finally spin-down.

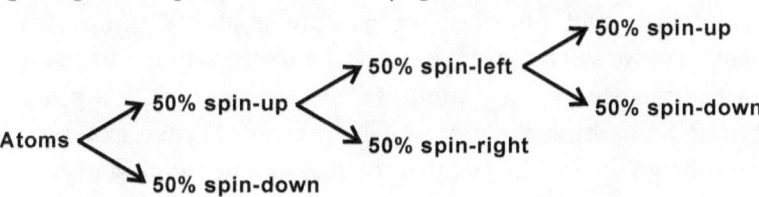

Fig. 16. When several Stern–Gerlachs are arranged in a series, but each apparatus is oriented perpendicular to the previous one, a spin-up atom can subsequently be measured as spin-down. Each perpendicular measurement effectively erases the previous measurement result from the world.

Approaching this from a common-sense mindset, one would assume each atom can have only one spin at any given time. Perhaps each apparatus is simply rotating the atoms 90 degrees one way or another. Again, varying the experiment rules out this explanation: If you prepare a stream of spin-up atoms, and then you pass them through a second apparatus that's at a slightly different orientation, it doesn't rotate the atoms by just that small amount, producing all "spin-up-ish" atoms. Instead, you'll get a statistical distribution of mostly spin-up-ish and a few spin-down-ish atoms, the probability distribution depending exactly and predictably upon the orientation angle you chose. Perhaps most tellingly, we never catch atoms in the act of flipping over and winding up on the screen somewhere between the two spots. We always get just the two spots.

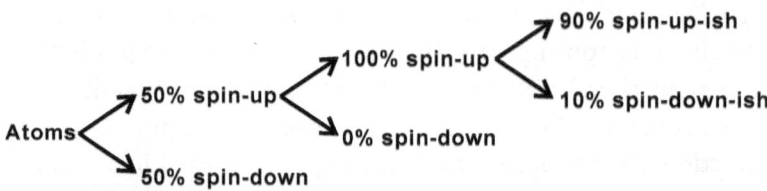

Fig. 17. Re-measuring spin along the same axis (middle) produces the same results. Subsequently changing the apparatus orientation slightly produces a distribution of results that's statistically predictable by the new angle.

All of this evidence seems to refute the hypothesis that the atoms are like little tops, each carrying with it information about its intrinsic spin. So, when we measure an atom's spin, in what sense can we say that we are collecting or recording information that pre-existed in the atom, like the information in a printed book? Something far more alien to our everyday experience must be going on. The fact that the measurements are statistically predictable, to a high degree of precision, should be a clue.

For any individual atom, the spin orientation we measure may depend on spin measurements we performed previously. If we haven't measured spin at all, the result is always randomly split

50/50. If we measure spin, and then we re-measure along the same axis, the result is 100/0 — all of the atoms will be measured the same as previously. A physicist might say that the first measurement has constrained the result of the second. But if we measure along a perpendicular axis (e.g., measuring left–right spin of an atom that was previously measured as spin-up), the result is 50/50 — there is no correlation between the measurements, and the second measurement result is not constrained at all by the first. In fact, the first measurement has been "quantum-erased" from the world. If we measure at an intermediate angle, the result will be statistically intermediate: The first measurement *partially* constrains the result of the second measurement.

Here we see an example of information constraining other information, something that will become important in Chapters 4 and 5. Information constraining information is seen most clearly in the phenomenon of quantum entanglement. It's the most compelling case that we live in a wiki-world rather than a book-world.

Information Transcending Space & Time

Now that you're familiar with the weirdness of intrinsic spin, you can fully appreciate the weirdness of entanglement. Particle physics tells us that in some types of radioactive decay, the nucleus of an atom spits out two of the same particle, for example, two electrons. When this happens, the particles share unusual qualities and are said to be *entangled*. They're a little like telepathic twins, rather than distinct individuals. If you do something to one electron, the other will seem to feel it — no matter how far away it is. Einstein was troubled by this prediction of quantum mechanics, calling it "spooky action at a distance." Since that time, sophisticated experiments have confirmed that entanglement does occur, and there are even hot technologies, such as quantum cryptography, that exploit the effect. However, interpreting entanglement, or figuring out what's really going on between two entangled partners, remains a topic of debate.

Consider a pair of protons (positively charged subatomic particles) ejected from an atomic nucleus. Each proton has an intrinsic spin. But physics tells us those spins must be opposite: The angular momentum of each proton must cancel out the other's. That's because if they didn't, the system of atom-plus-ejected-protons would have more total angular momentum (think: more energy) after the emission than before, and that can't happen. Newton's famous Third Law of Motion, that every action has an equal and opposite reaction, applies: The ejection of a proton with one spin must be balanced by the ejection of a proton with the opposite spin, so that the total angular momentum is conserved. Things always obey the laws of physics — even though sometimes, the results can seem bizarre to us.

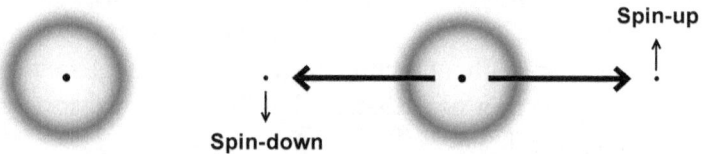

Fig. 18. If two protons are ejected from an atom's nucleus, the system's total angular momentum before (left) must equal the total angular momentum afterward (right). Therefore, the protons' spin must be opposite.

Since the two protons are entangled, this results in some extraordinary effects. If we measure the spin of the proton ejected to the right and find it to be spin-up, then we know for certain that the proton ejected to the left will be found to be spin-down. This happens 100% of the time, so the protons are said to be 100% *anti-correlated* with regard to spin. If we measure the left proton along a different axis, and perhaps find it to be spin-north, then we know for certain that the right proton will be measured as spin-south.

Already this is extremely peculiar, and we haven't even gotten to the variations. It's strange enough that we can measure the spin of an individual particle, and regardless of the orientation angle that we choose, we'll get a definite and unambiguous "either/or" answer, but never anything in between. Doing the same thing with entangled partners produces *two* either/or answers, and these answers are *always opposite*. Even if the particles are far apart — one experiment separated them by some seven miles — it's tempting to imagine that they are signaling each other about what measurements have been done, much faster than light, which as far as we know is impossible. It's no wonder new-age and parapsychology enthusiasts have embraced entanglement as some kind of spiritual effect transcending space and time. I won't go there, although there is little doubt that entanglement does transcend distances in space, at least in the physical sense.

When many people first learn of entanglement, they often come up with an intuitive explanation, one that's in line with the intuitive, book-world picture of pre-existing information: The particles are ejected from the atom merely bearing opposite spins from the beginning. Naturally in that case, if you measure one spin, you'll find the opposite spin in its partner, for the same reason that if you find a left shoe, you know that if you find its partner, it will be a right shoe. But this doesn't account for the fact that any orientation can be chosen for the spin measurements. If one particle is ejected that "really" is spin-up, and the other is "really" spin-down, then there's no explaining why those particles could also be measured as spin-left vs. spin-right, or spin-north vs. spin-south, or whatever. It seems that in order to account for all possible measured values of angular momentum, as in the book-world picture, a particle would have to carry an infinite quantity of information — one value in case it's measured this way, another value in case it's measured that way, with all opposite numbers embedded in its partner.

One revealing thing about entangled particles: You don't have to measure both of them along the same axis. Suppose you measure particle A to be spin-up. We know that if we then measured its partner (particle B) in the up–down direction, we would find it to be spin-down. But, what about if we measured particle B at a perpendicular orientation, such as spin-left vs. spin-right? Would you care to guess what the result might be? Particle B has a 50/50 chance of being either spin-left or spin-right. You can repeat this with as many particles as you like, and you'll always get the same result. There is no correlation between the up vs. down and left vs. right measurements — which makes mathematical sense, since those axes are perpendicular.

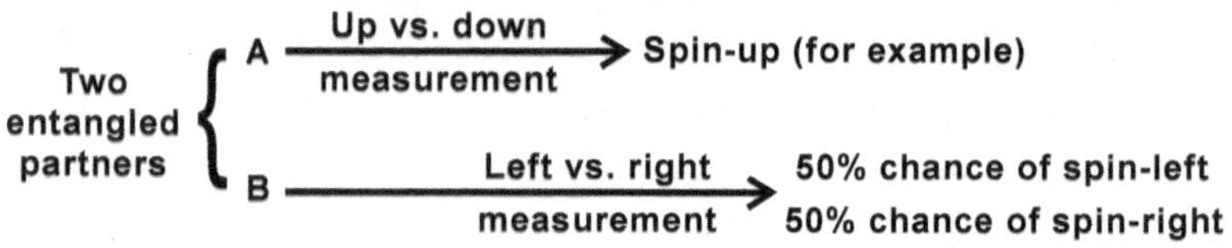

Two
entangled
partners
{
A ⎯⎯ **Up vs. down** **measurement** ⟶ **Spin-up (for example)**

B ⎯⎯⎯⎯ **Left vs. right** **measurement** ⟶ **50% chance of spin-left** **50% chance of spin-right**

Fig. 19. If entangled particles are measured for spin along perpendicular directions, the results will be completely uncorrelated (i.e., random).

Finally — and this could be my single favorite experimental setup of all time — we measure particle A to be spin-up, and then we measure its partner at a diagonal angle, like 10:30 vs. 4:30 on the face of a clock. Will particle B be spin-up-left. or spin-down-right? In this case, we find the particles to be *partially* anti-correlated. Measure a lot of pairs in this way, and *most* of the partners of the spin-up particles will be found to be spin-down-right. And, you can calculate that statistical distribution with easy math. Summing up: Measuring spin along the same direction = 100% anti-correlation; measuring spin along slightly different directions = partial anti-correlation; and measuring spin along perpendicular directions = 0% correlation.

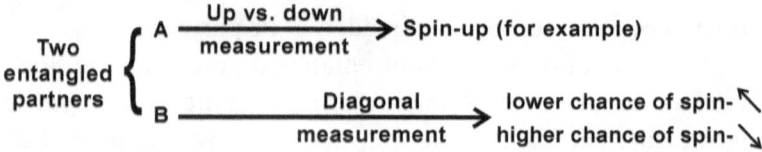

Two
entangled
partners
{
A ⎯⎯ **Up vs. down** **measurement** ⟶ **Spin-up (for example)**

B ⎯⎯⎯⎯ **Diagonal** **measurement** ⟶ **lower chance of spin-↖** **higher chance of spin-↘**

Fig. 20. If entangled particles are measured for spin along directions that are different but not perpendicular, the results will be partially anti-correlated.

These findings are incompatible with a book-world picture. It seems that particles of stuff would have to possess massive quantities of information for these experiments to turn out as predictably as they do. If nature can accomplish this feat some other way, which doesn't require extravagant redundancy and/or complexity, then we should take that explanation seriously. One such explanation is that our universe is a wiki-world.

Information on Demand

In *Physics and Philosophy* (1958), Werner Heisenberg wrote, "We have to remember that what we observe is not nature itself, but nature exposed to our method of questioning." Observing nature itself would be like opening a book and reading its contents. I claim that exposing nature to our method of questioning is like exposing a computer-generated wiki to our questions: When we seek new information, the information appears. However, any answer that's provided may be constrained by information that already exists. To examine this analogy, let's look at an example.

Our fictional *Reign of the Mantelopes* world involves a cast of characters, each of which is described in the *Mantelopes* wiki. Suppose there's a character who is half-man, half-antelope. Either his mother or his father could be the human parent, but we suspect that he got his human-ness from his mother. So we ask the computer whether or not his mother is human: Yes or no? Unfortunately, the computer has no other information that might help it decide. So, it chooses an answer at random, 50-50. Seeing the answer YES on the screen, and not knowing how the process works, we might be tempted to think, "The computer searched the database for a pre-existing answer to my question, it found the answer, and then it put that answer on the screen." In fact, the answer did not exist before we asked. The human-ness or antelope-ness of the character's mother was uncertain; there was equal potential for this question to go either way. Such a scheme simplifies the wiki's database considerably. Rather than containing the answer to any question that might ever be asked, the database only contains answers to questions that have already been asked. It operates on a "need to know" basis: Information is not a feature of the database until someone wants to know, and asks. This greatly reduces the amount of information that the database needs to hold.

Now, imagine instead that someone had already asked the wiki about the *father*, and as a result, the wiki had an entry saying that the father is human. That changes things; options are suddenly limited. The fact that this information is in the database will constrain the new information we seek. In order for *Reign of the Mantelopes* to be a logically consistent world,[4] when we ask the wiki whether the character's mother is a human, the wiki must now answer "No." Clearly, she must be an antelope.

According to the wiki-world picture of the universe, this is what's happening in the experiments: When we measure the spin of a particle that has never been measured, we're essentially asking nature a yes-or-no question — spin up or down, left or right, etc. — which has an equal probability of going either way. It's a perfect coin-flip, on the quantum scale. If we ask the same question again and measure the particle along the same axis, we will get the same answer. The old information constrains the new information; it's like looking two times at a coin that's been flipped only once. If we ask a *totally different* question, for example by measuring spin along a perpendicular axis, the old information does not constrain the answer at all, and again it's a random coin-flip. Rather than the particle revealing layer after layer of pre-existing information encoded in its stuff-ness, the information is appearing on demand. It's as if a wiki-of-the-universe is rewriting the entry on that particle, every time we make a perpendicular measurement of it.

When we measure entangled particles, it's even more interesting. The two entangled partners are quite literally like

---

4 Ignoring the problems associated with humans and antelopes interbreeding.

two sides of the same coin. When we measure the spin of one of them, it's as if we're doing a quantum coin-flip and seeing "heads-up": When that happens, it logically means that if you looked at the coin's other surface, you'd have to see "tails-down." With entangled particles, it doesn't matter whether or not you've asked about B — if you've asked about A, the answer for B is already determined. Information about A, being an existing feature of the world, constrains any information we might subsequently seek about B. The world would not be logically consistent if both A and B could possibly both be measured to be spin-up, for example, so they are guaranteed to be anti-correlated (i.e., opposite), 100% of the time.

Meanwhile, in the book-world picture of the universe, there is a struggle to interpret entanglement. One approach explains that the entangled partners carry anti-correlated information that is released when we measure either particle, or both. These are the so-called *hidden-variable* interpretations. An entangled particle, upon being measured, seems to signal to its twin what the answer to any subsequent question should be, even though this happens faster than light, in so-called *nonlocal* interactions.[5] Regardless of the interpretation, the book-world picture requires, at the very least, the specification of certain properties of everything in the vast volume of the universe. All of the associated particles — and we're talking numbers in the ballpark of one followed by 80 zeroes — presumably came into physical being at some point, each particle carrying intrinsic information such as mass and charge that identifies what kind of particle it is, but perhaps also additional information such as its momentum relative to other things. This *stuff* view assumes that information is a property of every particle in the universe, as if imprinted from on high by an omnipotent deity on the Day of Creation.

---

5 These interpretations don't actually claim that information is transferred from one point in space to another faster than light, but rather, that anti-correlations can stretch across distances in a "spacelike" or acausal manner.

If there were a way by which information could appear in the world only when something seeks out that information, then the universe could be much simpler than the book-world picture would require. If there didn't have to be all of those particles of *stuff*, the universe would be much simpler still. If information about things all over the universe were somehow all in one place, as it were, then the paradoxes of entangled partners seeming to signal across vast distances would disappear. Particles that collapse out of waves and probability distributions would reveal themselves to be more like projections or snapshots from a deeper, simpler informational world, like the shadows on the wall of Plato's cave. By the reality equivalence principle from Chapter 1, we would not be able to discern the difference.

Still, we cling to the book-world of stuff. As the physicist Dieter Zeh puts it, "Tradition seems to be incredibly strong even in the absence of any arguments supporting it."

Re-Thinking Observation, the Past & the Future

Ever since quantum theory was introduced, physicists have been trying to figure out what exactly is meant by "observation." Historically, the meaning was straight out of the book-world picture: Observation simply means deriving information from whatever it is we're observing, based on the information inherent in that object. Quantum physics made people confront the fact that the act of observing seems to have an effect on the system being observed. Contrary to some new-age claims, though, it's not that our minds are telepathically changing the physical world; after all, merely reconfiguring a particular experimental apparatus can result in something being observed and therefore changing the system on some level, as we see in cases such as the delayed-choice experiment. But, physicists cannot agree on the answers to certain questions. When does the observation process begin and end? What is physically happening in an observation? You can appreciate why these things would be problematic in the book-world picture of the universe: Reading a book or placing it on a flatbed scanner should not change the words printed on the pages. But that's kind of what seems to happen in an observation of a quantum system.

These kinds of difficulties disappear if you don't assume that the world is ultimately made of stuff, with information "printed" on it the way information is printed in a book. Instead, an observation is any event in which *new* information enters the world, or old information is confirmed. In this wiki-world-type picture, when situations occur that call for information to appear, such as setting up a Stern–Gerlach apparatus and then querying nature by measuring an atom's spin, then information is generated according to probability, like an idealized coin-flip. And with each observation event, the information becomes a feature of the world. In this way, the universe is constantly growing more complex. If we extrapolate this trend backward, we find that the universe started out much simpler, in informational terms, than it is today. As we'll see in Chapter 7, none of the information familiar to us was a feature of the world at that time.

Ah, time. Like measurement, the true nature of time has been the subject of fierce debate. The most common question (besides "What *is* time, anyway?") has been, why does time have an arrow, which points from the past toward the future? Every person experiences the sense of time flowing in one direction; we remember the past, but we don't remember the future. Why is this the case? After all, the laws of physics are time-symmetric — everything can happen in either direction. The equations of gravity, for example, describe how an apple should fall: Not only should the apple get progressively closer to the ground, it should also accelerate while doing so. However, the same equations also describe what an apple should do when it's tossed upward: It should get *farther* from the ground while also *slowing down*. Run that movie backward and the result will look just like the first case, an apple falling. And, a movie of an apple rising and then falling looks identical both backward and forward. This is how it is with virtually any law of physics — the direction of time doesn't matter.

Currently, physicists' best explanation for the arrow of time is the tendency of systems to move away from order and toward disorder, referred to as an increase in *entropy*. This process runs in only one direction: You can scramble an egg easily, but it's difficult to unscramble it. Entropy is well understood to be a simple matter of probability. For example, there's one and only one way in which billiard balls can be lined up in order along the edge of a pool table — but there are many, many ways that they can be scattered across the table. If the balls are bouncing around, at any given moment it's extremely unlikely that they would spontaneously regroup into that original one-of-a-kind arrangement, when there are many more arrangements they can be in. Thus there is a purely statistical reason for the laws of probability pointing in a direction of increasing disorder: Disorder is simply more likely to arise out of order than vice versa. This is why the arrow of time is sometimes said to *emerge* from entropy, or from the statistical, probability-driven nature of the universe.

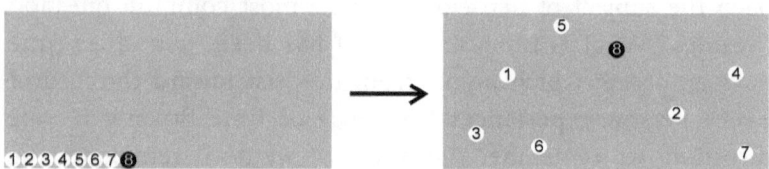

Fig. 21. There is only one way to arrange billiard balls at one corner of the table (left), but many ways in which they can be arranged otherwise (right, for example). It is therefore very unlikely for the arrow to reverse direction.

In a wiki-world kind of universe, the appearance of new information creates a fundamental distinction between the past and the future. In fact, it can define those terms altogether — while satisfying our conviction that there's more to the asymmetry between the past and future than only a matter of entropy. To illustrate this idea, I use the analogy of a movie theater.

When you sit in a theater, the movie runs and you get information from the screen. But that's only half of what's going on. In the back of the theater, there's a projector with the rest of the movie. Suppose the film is halfway over. Intuitively, the first half of the film is in your past, and the second half is in your future. If you haven't seen the movie before, you have no idea what's on the footage yet to be projected; it's uncertain to you. A character that you watched grow up and get a job, a short while from now, might win the Nobel Prize, or they might die in a car accident; kind of as in life, you have no idea.[6] On the other hand, the first half of the movie is settled, and if you could rewind the film, you'd be able to predict exactly what is going to happen, at least up to the point where you hit the REWIND button. You have a lot of information about the first half of the movie, but you have zero information about the second half. And turning around won't help; it's dark back there. The best you can do is let the movie run and absorb the information as it comes.

In a wiki-world universe, what we call the past is simply the direction that information comes from. In the movie theater, that direction is an arrow pointing from the screen toward you. Information comes from the direction of the screen; nothing comes from the other direction. Since the screen is constantly bringing you new information, you might be tempted to think of the pictures on it as happening in the present, "right now" — but you'd be wrong. Since the screen is some distance away, whatever you're viewing at any given moment actually appeared on the screen a tiny fraction of a second earlier. That's true of the world: Anytime you look at something and collect information about it, that information is coming from the past. The speed of light makes it impossible for the here-and-now moment that we call the present to have any true, universal physical meaning, as

---

6 There is a weakness with this movie-theater analogy: Since a movie's information pre-exists on film or on a hard drive, this is closer to a book-world picture than a wiki-world picture. So, try to imagine a film that's generated spontaneously, while you're watching it — the simplest-case scenario would say that's similar to how reality works.

Einstein's relativity proved. Instead, we might want to think of the present as an abstract boundary or surface, separating the information-rich past from the uncertain future. This boundary is constantly moving in a direction toward the future, as information from the past builds up behind it, like rain accumulating in a bucket.

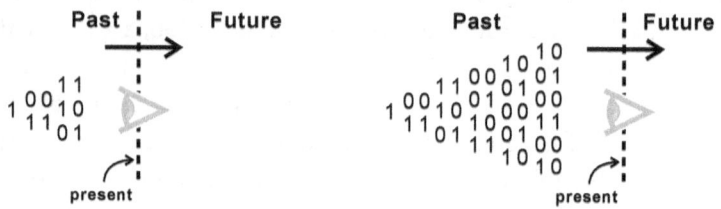

Fig. 22. As information accumulates in the past, we observers feel as if we are traveling in a direction toward the future, where the world is uncertain and devoid of information. This creates a natural arrow of time.

This conception of time eliminates the troublesome idea of retrocausality, which we saw in the delayed-choice and quantum-eraser experiments. We find retrocausality disturbing because, judging by the convention of the ever-ticking clock on the wall, it seems to let us go back and change the past, just by altering something in an apparatus setup, mid-experiment. It's difficult or impossible to square these experiments with the assumption that we live in a book-world of stuff, so the experiments, and quantum physics in general, become spooky and mysterious. However, if we allow that *new* information enters the world, and if we define the past as the direction in time from which that information comes, then the unease of retrocausality melts away. We aren't going back in time and altering the past in these cases, because those particular details of the past never actually happened! Even though the clock on the wall says that a photon must have encountered a double slit or a half-silvered mirror three nanoseconds ago, we can't really say that the event occurred in an ontological or fundamental sense, because no information about the event has appeared in the world.

This is obviously an unorthodox view of time, one that questions the assumption that time is a steady, uniform flow that marches ever-onward, affecting all of the world's stuff at once. There is no universal time-keeping clock against which all events in all places can be measured and described. But, we already knew that a universal time could not exist: It was vanquished over a century ago by Albert Einstein and special relativity. He showed that time, and even things like the simultaneity of two events, is relative to the observer — the past and future cannot be defined absolutely. We'll look closer at relativity in the next chapter.

A final note about time: If the universe is indeed informational, and the analysis above is accurate, then there can never be such a thing as time travel. Into the past, anyway — we are always, it seems, time-traveling into the future, as shown in **Fig. 22**. In fact, traveling into the future, by our new definition, is an unavoidable aspect of receiving any kind of information. When it happens, we lose a little piece of uncertainty in the world, and we move forward by one click on the ruler of time that way. But, by our new definition, information *always* comes from the past — so there's no way that a person could time-travel to 100 years ago and provide new information to the people there. For example, you could not go back in time and kill your grandfather (i.e., give him and the coroner information about his death), because this would contradict information about your own birth. The only way you could kill your grandfather would be if it were a murder–suicide, *and* you somehow managed to erase all traces of your influence as well as your grandfather's influence on the world, like in the quantum-eraser experiment. Now *that* would be an interesting time-travel movie . . . .

So, no time-travel, and no aliens? It turns out, these things might be mere products of our imagination, and like magic and leprechauns, they just cannot happen in a universe that's far simpler than we believe it to be.

Things to Remember From Chapter 2

• In addition to assuming that the world is ultimately made of stuff, Western science assumes that information in the world is derived from that stuff.

• Since stuff is assumed to be the ultimate ground of reality, and all of that stuff has been around since the beginning of the universe, it's assumed that at least some of the information in the world has been around for as long as stuff has been around.

• The results of certain quantum experiments, such as the delayed-choice experiment, are difficult to explain under the assumption that a particle possesses certain information as it traverses its path in the experiment.

• Instead, information may actually appear on demand: When we set up an experiment to obtain new information — an act that John Wheeler would call querying nature — the information did not pre-exist in the world. Instead, it only came into existence when we asked for it.

• The arrow of time emerges out of the appearance of new information. The past is defined as the arrow-direction from which information always arrives; the future is the opposite direction.

• A universe in which information appears gradually and minimally, on a "need-to-know" basis, is simpler than a universe that begins with a great deal of information about a lot of stuff.

CHAPTER 3

THE RELATIVITY
OF INFORMATION

I f the fundamental nature of things is information, and not some kind of *stuff* with information "written" on it like printed ink in a book, then we should take a closer look at information and how it works. In this chapter I'll show that information is not a standalone affair; information always represents some kind of *comparative* quantity. It is something that signifies either a difference/change, or an equivalence (i.e., the difference or change is zero). In addition, what we commonly call measurement is a compounding of information, in which new information that we receive is compared to old information that we already have. The concept of new information building upon old information explains how the world could gradually evolve to appear incredibly complex today, without complexity coming into being suddenly, in a mysterious, Creation-like event — the conventional *stuff* description of the Big Bang.

Measurement: A Relational Affair

The words "observation" and "measurement" are often used interchangeably. To measure an object's properties is to observe those properties. But, observation and measurement, strictly speaking, are not the same. If I see a thing out there in the world, that's an observation. A measurement, though, is a *quantitative assessment* of some physical aspect of the thing: how far away it is, how large it is, what wavelengths of light are reflecting off it (i.e., its color), and so on. In order to measure distance from myself, for example, I have to compare the observation of the thing's position with knowledge of my own position, and express that comparison in terms of some known standard of distance: "The thing is ten feet away from me." And that's what a measurement is: It's the process of comparing two observations (such as noticing that a thing out there in the world is some distance away from myself), and then *comparing that comparison* to an existing standard, such as the known length of a ruler. To say, "The thing is ten ruler-lengths away" is to express the distance in terms of the known length of one foot, to create a numerical relation, or a ratio — in this case, the ratio 10:1. When making a measurement, we are comparing new information with old information, and thereby creating a *compounded* form of information. This compounded information typically offers some meaning or application that we can use to our benefit, such as, "I need to walk ten feet in order to reach that thing over there." Lots more on this concept in the next chapter.

Measurement, at least in numerical terms, is not a natural process. Humans invented numerical measurement, when we were developing agriculture and transitioning from nomadic hunter–gatherer activities to more stable farming of domesticated plants and animals. People needed standard methods to mark off plots of land, and as they began specializing in particular crops and livestock, trading became important, and there needed to be standard units upon which trade could be based.

If you wanted to divide a plot of land into equal sub-plots, you needed to be able to measure length or distance. People discovered that you can do this with some kind of standard yardstick — a consistent unit of measure that could form the basis of other measurements. All you have to do is place the yardstick on the ground, and then move it so that the beginning is now where the end was, and you keep repeating this action along a straight line. If you want to measure the distance from here to there, count how many times you have to place the yardstick in order to get there, and the total distance is that many yardstick-lengths. If you want to set a distance that's a certain number of yardstick-lengths, just place the yardstick that many times. This can be done in any direction you like. If you start from one spot along a straight line and measure an equal distance in both directions, then when you're done, you know your starting point was the midpoint. Two such lines, measured equally and with their midpoints coinciding, create the diagonals of a rectangle, the edges of which form right angles at the corners. These right angles define what it means for two lines to be perpendicular, and if you make a rectangle's diagonals perpendicular as well as of equal length, you have a square. For a circle, connect all of the places on the ground that are an equal number of yardstick-lengths from a central point. And so forth. Geometry was born.

This process is so intuitive and familiar that we take it for granted, but it's actually kind of profound. In these examples of measurement (indeed in any example of measurement), we are establishing a numerical relation between two quantities. These quantities are abstract: One is the length of our chosen yardstick, and the other is the distance between the two points we are measuring with the yardstick. But the idea of distance is, itself, abstract. The only way you can measure the length of a yardstick is with another yardstick, or something like it. Measurements of distance are not based on anything absolute, or fixed in the natural world. In the early days of numerical measurement, anatomy-based units such as the cubit or foot were convenient if imprecise reference lengths, but even today, if you want to measure distance, you have to create that numerical relation with some kind of previously established reference distance. Creating such a relation is the only way a measurement can have any meaning at all.

Before 1960, the length of the modern standard meter was literally defined by a metal bar housed under specific conditions at an international institute in France. Before that, it was one-ten-millionth of the distance between the equator and the North

Pole. Each time the meter was redefined, it was an attempt not only to make the standard meter more precise, but also for meter measurements everywhere to be more consistent. Finally, in 1983, the meter was defined as it is today: It is the distance light travels in a vacuum in exactly 1/299,792,458th of a second. The second, in turn, is defined as exactly 9,192,631,770 oscillating transitions in an atom of the element cesium, at a temperature of absolute zero — as cold as anything can possibly get. So rather than the meter being based on something arbitrary and inexact, the meter is now rigorously defined according to the invariable qualities of a particular kind of atom. Thanks to the quantizing of the atom discussed in the previous chapter, we can define the meter and second in terms of an atomic event that never changes. At last, these units are based on something seemingly absolute, as if God himself declared exactly how long a meter and second should be. (Actually it was scientists with good technology.)

However, despite the astronomically precise modern definitions of certain measurement *units*, the fact remains that all practical measurements are numerical relations involving these units. The duration of a typical TV commercial can be expressed as a relation with the standard duration of time, the second: "The commercial is 30 seconds in duration." Just as an early farmer may have measured a plot of land as being 30 of his standard yardstick-lengths on each side, you can measure a commercial as being 30 standard duration units. And you could express the duration of one minute as being 60 seconds, or two commercial-durations. If, however, you wanted to express the duration between transitions of a cesium atom, the best you could do is say something like it's 1/9,192,631,770th of a second. This is a chicken-or-egg problem; whenever you want to express a quantity that carries any meaning or significance at all, it *always* has to be expressed as a relation involving some other measurement.

Mind you, I am talking about numerical measurements carried out with some precision, which as far as we know is exclusive to the human intellect. But the act of measurement, in some sense, is also carried out by animals and even plants. A plant might start sending out flowers when it determines that the days are growing shorter. What defines "short"? That's hard to answer, but this threshold must be encoded into the plant's genetics somehow; after all, a different kind of plant might flower when the days are lengthening. Somehow, via the changes in hormone production or what have you, the plant is able to compare day-lengths across the weeks and months, and it responds accordingly. This may not be a precise numerical measurement such as those that humans make, but it is a measurement nonetheless.

At this point, you might be asking what all of this has to do with the universe. That will become clearer as the book progresses, but for now we're just examining how information relates to other information, a key concept in the simplest-case scenario. The important thing to realize is that measurements, descriptions, and information in general are never absolute. In order to have any practical usefulness — what we humans call "meaning" — information must always relate to other information in some manner. But, new information relating to other (presumably older) information is not something that could have gone on for an infinitely long time in the past; it had to start sometime. According to this book, that sometime was an extremely simple beginning to the universe, with complexity evolving for billions of years ever since, eventually to produce you, me, and an astonishingly rich Cosmos that seems to stretch infinitely in every direction and begins with a Big Bang.

In order to get into the proper mindset for this challenging idea, let's look at examples of information in everyday life, and see how it always represents some kind of relationship.

The Relativity of Position, Time & Velocity

There is no way to state the absolute position of something. Anytime we want to indicate something's position, we can do it only in reference to other things. And, those things can be described only in reference to other things still, or to the original thing.

If you ask me where your keys are, I'll have to say something like, "They're on the right side of your desk." I cannot answer that question with a set of numbers, like a cosmic latitude and longitude, that unambiguously pinpoints your keys' absolute location in space. If the universe were empty except for your keys, describing their position would be not only impossible, it would be completely meaningless.

In the 1600s, René Descartes invented the modern method we use to map out positions and distances. Today the system is called *Cartesian coordinates*, and it's the familiar graph framework you learned in math. Cartesian coordinates plot the position of any point on a line, or in a plane, or in a space — or in a system with more than three dimensions. In one dimension, the Cartesian system is a number line. In two, it's a graph with an x-axis and a y-axis, usually plotted horizontally and vertically. In three dimensions, a z-axis is added. On the surface, one might think that Cartesian coordinates provide the framework to specify the absolute position of anything, but there are three issues. First, where in space do we place the origin, that point marked "0" where the axes converge? It could be anywhere; it's completely arbitrary. You might choose to use the front-left corner of your desk as the origin, or your mailbox, or the tip of the Washington Monument. Second, the orientation of the axes is arbitrary. Even if we agree to place the origin at the front-left corner of the desk, should the x-axis be parallel with the desk's front edge? Or parallel with one of its legs, or with some diagonal line? It's an arbitrary choice. Finally, what is the scale we are using to mark off distance units on each axis? It could be inches, or centimeters, or any arbitrary unit. So, we need to specify three things[1] — an origin, an orientation of the axes, and a unit of scale —

---

1 There's one other parameter, handedness or chirality. An xyz coordinate system and its mirror image are not identical. There is a convention in science of preferring a "right-handed" coordinate system — for example, the angular-momentum vector illustrated on page 52 points upward. Curl your right-hand fingers in the direction of the wheel's rotation: Your thumb will point up, the direction of the vector specified by this "right-hand rule."

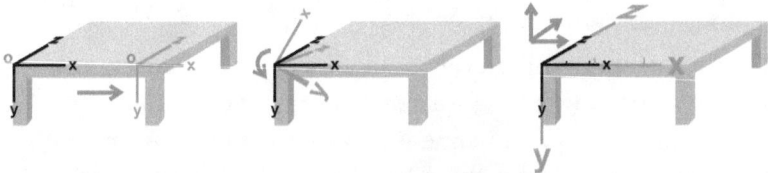

Fig. 1. Three things must be specified to establish a frame of reference: point of origin (left), orientation (center), and scale (right). These make it possible to create meaningful descriptions of position, rotation, and expansion/contraction, respectively, relative to this reference frame.

in order for the coordinates to be useful in any way for indicating location. A coordinate system with all of these parameters specified establishes a *frame of reference* in which any measurements might be taken. The term "reference frame" comes up again and again in physics, and also in this book.

Suppose you've specified a system of coordinates for describing the location of your keys: The origin is at the front-left corner of your desk, the x-axis is the front edge, the y-axis is a line pointing from the origin to the floor, and the z-axis is the left edge. For a standard unit of length, you've chosen one foot. The position of your keys within this reference frame, then, can be specified with three numbers: the distance, in feet, between the origin and your keys, as measured along the axes x, y, and z. It might be something like (2, 0, 0.5): This specifies that you would reach the keys if you started at the origin and traveled two feet to the right, zero feet up or down, and half a foot toward the back edge. Using a different reference frame, or just specifying a different origin or unit of length, will likely result in completely different numbers specifying the same point in space. This is why establishing a reference frame is critical in any measurement of position.

When we talk about time, and describe the time of some event happening, we encounter the same complications that we have when describing position. For this reason (and others), physicists often describe time as being like a fourth dimension of space, represented by a t-axis. Like position, we can't specify an absolute time;[2] the best we can do is specify an origin (like the time of some other event) as well as a unit of duration, such as the second. (We don't have to worry about orientation, as there appears to be only one dimension of time.) Along with a description of position, this allows us to describe the location, in space *and* time, of any event. For example, if you set the origin point of time as your alarm clock going off, we might say that you found your keys at the (x, y, z, t) "position" of (2, 0, 0.5, 500), which means: near the back-right corner of your desk as before, and additionally, 500 seconds after your alarm clock went off. As with position, we need to specify some point in time as a reference, from which other points in time can be measured — a distance in time between the two points, if you will.

None of that should be very surprising. But now let's combine distances in space with durations in time. If something is observed at one point in space at one time, and then it's observed at another point in space at another time, we say that the thing's position changed, or that the thing moved. We can measure the difference in position (i.e., the distance between the two points in space), and we can also measure the difference in time (i.e., the duration between the two points in time), and divide to calculate the velocity. A car traveling along a number line marked off in units of feet, starting at 0 at noon and ending at 5,280 at exactly 12:01, is said to have a velocity along the number line of one mile per minute or 60 miles per hour. But that velocity description is valid only in reference to the number line we specified. A different reference frame might have the number line pointing in the opposite direction (see **Fig. 2**), where the car ends up at −5,280; in that case, we'd have to say that in this reference frame, the velocity is −60 miles per hour. We can even establish a *moving*

2 Arguably, an absolute time could be established by the very beginning of the universe, the Big Bang as described in modern physics happening 13.8 billion years ago. But even that has difficulties. According to a theory by David Wiltshire, the age of the universe differs from place to place, due to the cumulative effects of general relativity.

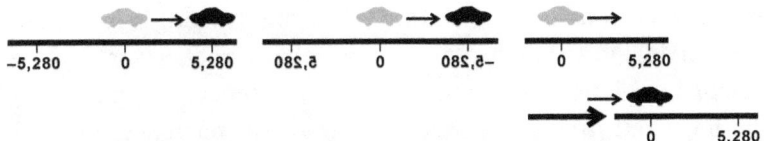

Fig. 2. A car with the same motion — depending on the choice of reference frame — may be described as traveling in the positive direction (left), the negative direction (center), or not moving at all, if the reference frame itself is moving at the same rate. That rate, in turn, would have to be described relative to something else yet, such as this page.

reference frame: If we've chosen another car moving along the road as our origin, and it's moving at the same rate, then relative to that car, the first car isn't moving at all. Physicists say that like position and time, velocity is *relative*: It is impossible to describe a velocity in absolute or independent terms. Velocity always has to be described relative to some frame of reference. This is known as the *principle of relativity*.

Most people associate relativity with Einstein, but actually, Galileo discovered it, in 1632. The Galilean principle of relativity simply says that in any given reference frame, the laws of physics operate in the same predictable manner. This means that as long as we are not accelerating,[3] there is no way to tell whether we are moving or not. If we are in a windowless spaceship, we can conduct experiments all day to try to figure it out whether we're moving, but the results will always be the same, and we won't get an answer. You should be able to appreciate this, given that right now, you are on the surface of the Earth which is rotating, and the Earth is moving around the Sun at some 67,000 miles per hour — but you can't tell.[4] The same would be true of a very smooth jet flight at cruising altitude. (The bumps of turbulence

3 The case of accelerated reference frames was not solved until Einstein's general theory of relativity in 1915.
4 Since the Earth is rotating as well as orbiting the Sun, there are slight accelerations that fluctuate daily, and these could be measured by sufficiently sensitive instruments. But except as manifested in things like the tides, the tugs that these accelerations produce are imperceptible, and it feels as if we are standing still.

are little periods of acceleration.) For this reason, it is literally meaningless to say that something is either stationary or moving, *except* in relation to some frame of reference. Physicists use the term *rest frame* to describe the frame of reference in which something appears to be at rest, relative to the observer. If something is in the rest frame, for all intents and purposes it's standing still, and all measurements you make will reflect this.

It's interesting that the world does not come supplied with some kind of eternal framework of absolute reference rulers, with an absolute clock that is valid for all observers. Even if our universe is a book-world (Chapter 2), then the information carried by a thing made of *stuff* — at least with regard to position, time, and velocity — can have no value except as compared to those properties in other things. The three qualities of position, time, and velocity are only describable relative to the position and velocity of other objects and the time of other events; there is no escaping that fact. This is why, when describing things like velocity, the word "observer" invariably pops up. It's convenient to say that according to an observer on the ground, a jet is going 500 miles per hour; the speed might be described differently by other observers in other jets. The observer concept is also central in Einstein's relativity, which we'll see in a moment. It's not a coincidence that observer language is used in quantum mechanics as well, as we'll see in Chapter 5. These ideas are all connected, and that's what this book is about.

The Relativity of Space vs. Time
Information regarding position and velocity are relative, and information about time is relative. Where things get seriously interesting is that position, time, and velocity are all *relative to each other*. This was one of Einstein's greatest insights, and he called it the *special theory of relativity*, or just special relativity for short. It's special because it's a special case of a more general theory that he put together later (the general theory of relativity). Einstein's relativity concepts scare off a lot of people, who maybe

have convinced themselves that they could never understand them, but that's unwarranted. It's not that difficult to understand special relativity, especially if you have some analogies to help.

Special relativity refers to the fact that space and time, or better said distance and duration, can trade off with each other. To grasp this idea, think of distance and duration as two different, but related, kinds of *extension*, or *separation*. If something is extended in space, for example two points are separated from each other tape-measure-wise, then we refer to extension as distance or length. In time, we refer to extension as duration. Now for the analogy: Think of distance and duration as two different "currencies," like dollars and yen, that can both "pay" for the same thing. Amazingly, special relativity shows that these quantities can be exchanged, like money. An amount of one kind of extension is equivalent to a certain amount of the other kind; for example, a length in time (duration) is equivalent to a particular length in space (distance). And, we actually know the exchange rate between distance and duration. It's a conversion factor, just like a dollars-to-yen number you might see in a bank or scrolling across the bottom of CNBC. That exchange rate is something familiar to you: It's the speed of light, a constant 186,000 miles per second in a vacuum. What that means is, one second of duration in time is equivalent to 186,000 miles of distance in space.

This isn't just a metaphor. When something is observed to travel through space, it is also observed to pass through time a little slower than otherwise — something that has been tested with atomic clocks aboard jet aircraft. Just as a lot of yen can convert into a few dollars, moving through a lot of space can convert into some time lost. In fact, a similar effect is what causes an object to fall in a gravitational field, where spacetime is curved:[5] Its travel through time literally starts to get converted

5 To see this demonstrated mechanically, search online for my video called "How Gravity Makes Things Fall."

into travel through space — relative to the observer of the falling object, anyway. And, as it continues to travel through space at a faster rate, it travels forward through time at a slower rate.

We can still trade between moving through space and moving through time, even without anything falling. If you fly past me in a rocket ship, you and I will not measure space and time the same. Judged from my reference frame, you are moving through space faster than I am, so I will see you moving through time *slower* than I'm moving through time. The same is true in your reference frame: You appear to yourself to be stationary, while I appear to be moving fast by. If we compared clocks, we would both find our own clock to be ticking faster than the other clock. Furthermore, we would each measure the length of the other person's rocket ship to be *shorter* than we would in the ship's rest frame — this is the well-known effect of relativity known as Lorentz contraction. The ultimate conclusion of Lorentz contraction, and special relativity overall, is that if you could view the universe from the perspective of something moving at the speed of light, all distances and durations, of everything in the world, would contract to nothing. The universe would appear to be a point with no extension whatsoever, either in space or in time.[6]

These ideas are, of course, inconsistent with the idea of absolute measurement, and of absolute information. The measurable separation between two things in space, or between two events in time, depends upon the observer's perspective. There is no preferred reference frame in which we can claim that something is "really" ten feet long, because in another reference frame, that distance might be measured to be nine feet long. Furthermore, no one can claim that such an object has "really" been there for ten seconds, because in another reference frame, that duration might be measured to be 11 seconds.

---

6 This is an intuitive way to understand why nothing can go faster than light — distances and durations cannot be measured to be any shorter than zero.

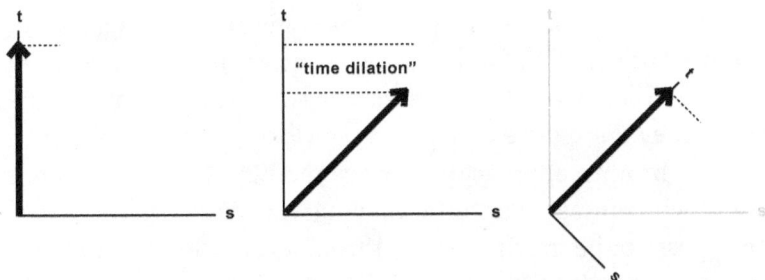

Fig. 3. The relativity between space and time can be understood in a simple geometrical manner. In a given reference frame, an object can be traveling in time but otherwise standing still in space (left). If it travels through space as well, then it necessarily travels less through time (center). In a reference frame moving alongside the object, however, the object would once again be described as stationary (right), and no time dilation would be observed.

Informational descriptions aren't rigid and single-ended, like a yardstick bolted to the ground; they are flexible, more like an accordion. If you're noticing similarities with the uncertainty principle (see page 44), where the values of measurements can trade off and depend upon the way we measure them, you aren't imagining things. All of these concepts are connected, and that's what makes physics so endlessly fascinating.

The Relativity of Mass

I was interested in relativity as a kid, but one of its claims seemed strange — that an object gains mass as it goes faster. How could something gain mass? It didn't make sense. Since then, I've learned that in a universe with a constant speed of light, observed mass must also be a relative quantity. Like distance and duration, the amount of mass that you would measure in a thing depends upon the reference frame in which you make the measurement. Even information that we normally think of as the *amount* of stuff depends upon how we choose to measure it.

The physics writer Lewis Carroll Epstein has a fantastic thought experiment to demonstrate how mass must be relative to the observer. Imagine that two twins have been arguing.

They are on separate passenger trains going at a very high speed, relative to each other, and each twin knows that the other train is going in the opposite direction. The twins realize that at some point, they will pass each other very closely. Finally, each twin sees his brother approaching. But something is peculiar. Since the other twin appears to be moving very fast through space, he appears to be moving slower through time. So, they each get the same idea: It's the perfect chance to take out the other twin with one single punch. The other twin is slower now and can't possibly compete. So, Twin A winds up and takes a full-speed swing at his sluggish-appearing brother. At the exact same time, Twin B takes a swing at *his* sluggish-appearing brother. What

The Relativity of Energy

One quantity or property trading off with another, the way distance and duration do in special relativity, is surprisingly common in physics. You might have learned that an object such as a comet orbiting the Sun trades off one type of energy for another. When it's far from the Sun and moving slowly, it is said to have a high *gravitational potential energy* and a low *kinetic energy*, the energy of motion. As it falls toward the Sun, it loses potential energy while picking up speed. Some of the potential energy is converted, through the effect of gravity alone, into kinetic energy. When the object passes around the Sun and heads back toward deep space, it slows down and starts gaining potential energy again.

At all points along the object's trajectory, the potential energy plus the kinetic energy is a constant number, due to the law known as the conservation of energy. Think of it as being like the height off the ground of both ends of a see-saw: As one end goes up, the other goes down. If we added the heights, we'd find that they always come out to the same total. The same is true of the object's two kinds of energy. However, the value of this total is not an absolute number; like so many measurements will happen? The situation is the same for each twin, and the laws of physics operate the same for both, so who decks whom? There is only one way to resolve this: In each case, the opposite twin's fist must appear to gain mass, so that despite being slower, the momentum of the punch received will be the same as the punch delivered. Each twin finds that his brother's punching mass is greater (albeit slower) than his own, and since momentum is defined as mass times velocity, the numbers cancel each other out. Both twins knock each other out equally. Just as Einstein discovered that space and time must trade off like two different monetary currencies, he also realized that mass is literally in the equation as well (famously distilled into the mathematical

mentioned in this chapter, it depends upon the frame of reference. Judged from a distance, and described relative to the Sun (i.e., in the Sun's rest frame), the potential energy or kinetic energy grow large at either end of the orbit, so the total is always large. However, if you were in a spaceship falling along with the object, right next to it — i.e., in the *object's* rest frame — you might measure it to have zero kinetic energy all the while. Naturally, all of the above would still be true no matter what units you chose for the measurements, or whether you, the object, and the Sun were all moving uniformly in some direction, and so forth.

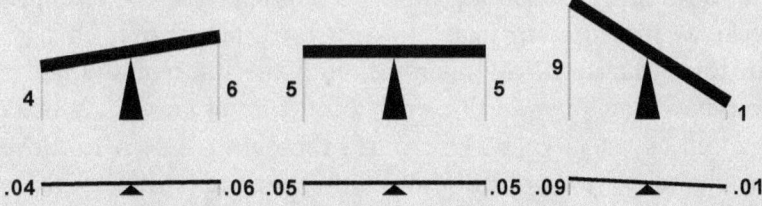

Fig. 4. The two ends of a see-saw (top row) maintain the same total height, like the total energy of a falling object. In a different frame of reference, however (simulated at bottom by squashing the diagrams), the total may be different — even though the relations do not change. This consistency is similar to the invariance under stretching that's illustrated on page 27.

expression $E = mc^2$). Mass depends upon the reference frame of the observer.

This kind of mass measurement, where relative velocity is factored in, is called *relativistic mass*. The more familiar variety is called *rest mass*, where we ignore this effect. (The difference is imperceptible in everyday life.) When you see a number indicating the mass of, say, an electron, that's its rest mass. But remember that nothing in the universe is truly at rest; to claim that something is absolutely at rest would be a meaningless statement, because there are always reference frames where the thing appears to be moving. "Rest mass" simply means mass that is measured in the object's rest frame. You could say you are at rest beside the object, or moving along at the same speed; it's the same situation regardless, just in different words. Choose any other reference frame — any speed or direction — and an object's observed mass, that piece of information as measured relative to you, would be different, were you to obtain that information. Mass, like position, distance, velocity, time, and duration, depends upon a relation between things and the perspective provided by that relation. Information about mass is relative.

Other Relative Measures

Position in space and time, and mass, are three properties by which stuff can be described. Many other familiar properties are derived by combining these basic properties. For example, velocity measures changes in position (distance) over changes in time (duration). Momentum, both the linear and angular varieties, combines all three: It's a function of an object's mass as well as velocity, i.e., how much the object is seen to move or rotate per unit of duration. Other measures that combine distance, duration, and mass include kinetic energy or work, force, pressure, and power. Since distance, duration, and mass all depend on the frame of reference in which they are measured, then all of the derived quantities do, as well.

Since relativistic tradeoffs rarely if ever enter our everyday experience, this may partly explain why we convince ourselves that things carry with them rigid, predefined information about their absolute properties. But imagine if we lived in a world where the speed of light was much slower, like 50 miles per hour. Just being out for a jog would seem to change the properties of the things we see in the world. In a world with light that slow — or, we could say, a world in which one second is equivalent to 73.3 feet instead of 186,000 miles — we would be intimately familiar with the relativity and flexibility of the information from the world around us. We might never even entertain the idea that objects have set properties or "carry" information; it would be natural, quite obvious even, that information depends upon the relations and relative motions among things in the world. But that's not how it is. Not until the 20th century did evidence appear showing that information about an object depends upon its motion relative to the observer seeking this information.

There are properties that do not, or at least may not, be measured differently depending upon relative motion. For example, the electric charge of a proton is always defined as +1, regardless of the reference frame. Temperature appears to be an absolute kind of scale; there's even an absolute zero at the bottom of it. However, opinions differ on whether temperature would be measured differently depending on the frame of reference, and presently there is no feasible way to settle that question with an experiment. Even in these cases, however, notice how we can't get away from describing properties in comparative terms: A proton's charge of +1 means one unit of charge in the plus "direction," as compared to a neutron or neutrino or photon. If everything in the world had the same charge of +1, the concept of charge would never be conceived, and charge wouldn't exist, in any sense. The same thing can be said about temperature; if nothing were warmer or cooler than other things — if everything were the exact same temperature — then the entire notion of temperature would be altogether whimsical and imaginary. You can experience this for yourself: Propose a property called "shemperature," and specify that according to you, every single thing in the universe, in all reference frames, has a shemperature of 42 shemps. You won't get very far with that proposal! To make any kind of meaningful description, we *always* have to compare new information to old information — just like the ancient farmer measuring a plot of land by using a standard measure of length.

Consider an imaginary universe containing one electron and nothing else whatsoever. Its position, velocity, charge, and any other property you can name would be meaningless. Mass, for example, is a property expressed (in the real world) in terms of gravitational interaction with other masses, so given no other masses, and no means by which to measure mass, the electron's mass would be undefinable. Even in a universe with *two* electrons, could the distance between them be described? Only in terms of some other recognizable unit of distance in that universe, and there is none. You might think the diameter of an electron could be a candidate, but physics treats electrons and other fundamental particles as points with no extension in space. Not until you consider *three* electrons can there be any meaningful talk of distance, and in that case, the best you can do is talk about ratios. If electron A and C are closer together than A and B, then we might specify AC as our unit of length, and measure that A and B are separated by 30 AC's. Arbitrarily we might instead choose BC as our unit of measurement to describe the distance between A and B. (See **Fig. 5.**)

No matter how you slice it, any information that specifies a property of a thing, or a measurement of a thing, is a two-ended affair. All information represents some kind of comparison or relative difference (or relative sameness) that is observed in the world. There is no such thing as absolute information; the term itself is an oxymoron. Try looking at it this way: If we took a single electron and placed it in a universe with nothing else in it,

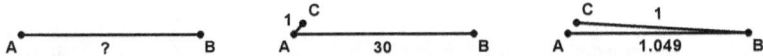

Fig. 5. In a universe containing only the point particles **A** and **B**, their separation could not be described relative to anything else (left). Adding a third particle, **C**, makes it possible for there to be a ratio of distances (center). The choice of unit specifying the amount of separation, like the choice of reference frame in **Fig. 2**, is arbitrary: **AC** could be chosen as the value "1" (center), or just as easily, **BC** could be the reference unit (right).

would it still be an electron? I don't mean that in a tree-falling-in-the-woods sense; how could we say it has charge or mass, if there's nothing for it to gravitate toward or attract/repel? Arguably, the very *existence* of an electron depends upon the presence of other things with which it can relate and interact. This is a subtle ontological point, but it's yet another reason to be suspicious of the assumption that the universe consists of absolute, persistent particles of stuff bearing any kind of absolute, persistent properties.

As summed up in a brilliant fortune cookie I once got, "If absolute can be defined, then absolute is not absolute anymore." Everything is defined relative to other things; it's just the way the world is. This is a *relational* universe through and through, and we should take this idea extremely seriously when contemplating the biggest questions mankind has ever asked.

Rich Information, Rich World

There's no question that the world is rich with information. From a fish glimpsing an approaching predator to a human reading a book, living organisms are awash with it, all telling us about our surroundings in one way or another. The information is so detailed, and the picture it provides of the world is so sharp, we humans take it for granted. We assume that the informational richness is merely a reflection of the extreme complexity of the stuff of the world, and as organisms, we must be extra-complex packages of this stuff, wandering about the planet in our own separate existences, searching for the meaning of it all. This view of the world as consisting of interacting but otherwise isolated particles of stuff, despite being intuitive, is a lonely and impoverished worldview. By the end of this book, you'll see how each of us is an information processing system that's intimately tied to every other living thing — and together, the tree of life that comprises all of us is tied to the farthest reaches of the Cosmos.

Things to Remember From Chapter 3

• When we make a measurement, we obtain information. But no measurement produces absolute information: Every measurement is a relation of some kind with older or previously obtained information. If we measure the height of a tree with a ruler, the number we get is a ratio that expresses a relation with the ruler-length we had defined earlier, as a reference.

• All measurement units are like this. There's no such thing as a measurement unit that isn't defined in terms of some other unit.

• Measurements of distance, time, velocity, mass, and energy are all related, and therefore they all depend upon each other. Any information about these properties has meaning only in relation to a chosen system of other measurements, called the *frame of reference*. Measurements made relative to one reference frame can be different from those in other reference frames.

• Information is relational and dependent upon other information. This complicates the assumption that the universe is ultimately made of stuff, each particle of which has certain absolute properties or values.

CHAPTER 4

SO WHAT EXACTLY
IS INFORMATION?

We've covered a lot of ground so far. Let's take a moment to review what we've discussed. In Chapter 1, we compared two possible views of the nature of things: one where the world is made of *stuff* plus information about that stuff, and another where it's just the information, with *stuff* emerging from that information. Chapter 2 elaborated on these two views by asking whether information in the world sits around like the words in a book waiting to be read, or whether it appears in the world on an ongoing basis, when some system seeks out information or "queries nature." I argued that not only would the latter kind of universe be far simpler, but the well-known experiments of quantum mechanics are perfectly consistent with that view. By contrast, quantum experiments are weird and extremely counterintuitive if we insist on the ironically "intuitive" assumption that nature consists fundamentally of stuff. This assumption is further weakened by the realization that there is no such thing as absolute information, discussed in Chapter 3. All information about things expresses some kind of relative difference, or change, or equality, or some other comparison. In this chapter, we'll take a close look at what information is, and how it does what it does. I'll try to demonstrate how the universe could be made of information *only*, with no persistent, absolute particles at all in the fundamental picture.

Information: The Reduction of Uncertainty

We've looked at the types of properties that information can represent — the numbers or quantities that information can tell us about a thing, such as its mass, velocity, and so on, none of which is absolute. But what makes a number or a quantity "information"? After all, 6.5 is a number, but is it information? If we could understand the nature of information itself, we would be in a much better position to understand how information may be the key to the nature of *things*. As a bonus, we'll see how information connects many of the concepts we've discussed up to this point, which is another clue that the nature of things is ultimately informational.

A good general definition of information comes from the anthropologist Terrence W. Deacon, based on insights by the mathematician Claude Shannon, who is considered the father of information theory. Deacon, whose research at U.C. Berkeley involves human evolution and neuroscience, broadly describes information as *the removal of that which would be statistically expected*. If you flip a coin many times, your statistical expectation is that half of the flips will come up as heads and half will come up as tails. If you discover this not to be the case — maybe heads comes up 60% of the time, in the sense that the more times you flip, the closer the percentage converges on this number — then you gain information that something funny is going on with the coin. Furthermore, if you hand the coin to someone, and you pass on this information, "Sixty percent of the time, the coin comes up as heads," then the information will alter the statistical expectations of the person who receives the message. Before the message, the receiver was expecting an even 50/50; afterward, that balance is tipped in one direction, as some of the original statistical expectation is removed — specifically, the expectation of 50% tails becomes an expectation of only 40% tails. The number is information because it has *constrained* the receiver's statistical expectations, resulting in new expectation numbers. That is what information does.

The physicist and theoretical biologist Bernd-Olaf Küppers approaches this same idea from a slightly different angle. He writes, "An important task for information is to eliminate or counteract uncertainty." In the example above, an absence of information about a coin makes the result of any individual coin flip completely uncertain; it could go either way, with equal probability. But if you receive information that the coin has a 60% chance of coming up as heads, then your uncertainty is counteracted somewhat. Furthermore, if we're talking about a coin flip that has already happened, then looking at the coin will *eliminate* your uncertainty altogether about that one flip. This is what information is and this is how it works: Starting with a statistical expectation that may be completely uncertain, information constrains that expectation to some degree, anywhere from just a little (counteracting uncertainty) to a lot (perhaps eliminating uncertainty completely).

So, suppose someone hands you a coin, and then tells you that the coin comes up as heads 60% of the time. Then you flip it once and see that it's tails. You're receiving information in steps, and in doing so you progress from total uncertainty (I have no idea how this flip will go), to reduced uncertainty (at least I know that it's slightly more likely to be heads), to complete certainty (actually, it was tails).

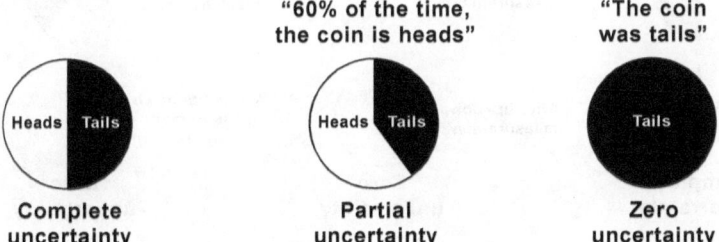

Fig. 1. Starting from complete uncertainty (left), a piece of information can reduce our uncertainty somewhat (center), at least in a statistical sense. Information specific to an individual event in the past can eliminate uncertainty about the event altogether (right).

Sometimes information reduces our uncertainty, and sometimes it totally eliminates our uncertainty. Think back to the Stern–Gerlach experiment (pages 53–57), where silver atoms are separated by spin. At the start of the experiment, measuring spin in the up–down direction, any individual atom may be measured as spin-up or spin-down; there is a 50–50 chance of either happening, and the result is completely uncertain. However, if you measure an atom's spin and find it to be spin-up, and *then* you measure the same atom in a diagonal direction, you will find that the information from the first measurement has partially constrained your statistical expectations of the second. Instead of being completely uncertain of how the results will turn out, as in the first measurement, you now have an idea: An atom that was measured as spin-up has a higher chance of subsequently being measured as spin-up-and-to-the-right. If you decide, however, to perform two up–down measurements in succession instead, then the first spin-up measurement has reduced the uncertainty of the second measurement all the way to zero. You are certain that the spin will still be measured in the same direction, because of the information already in the world about the atom.

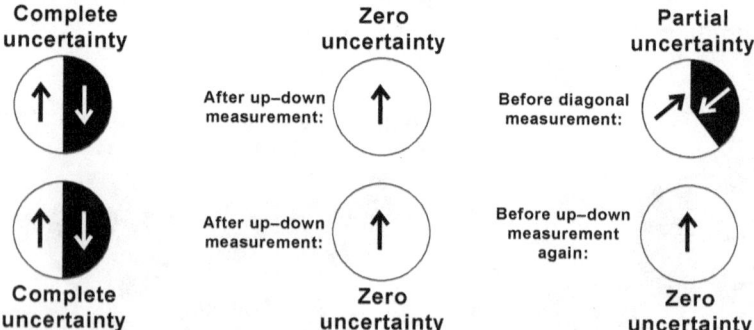

Fig. 2. An atom that is measured as spin-up subsequently has a higher chance of being measured as spin-up-and-to-the-right (top). Information from the first measurement constrains our expectations of the second measurement. The same is true if we simply repeat the measurement (bottom), although in that case our expectations are completely constrained.

Something interesting is going on here: Information can reduce our uncertainty of things that have not yet happened, like the probability of the outcome when we flip a trick coin, or if we measure the spin of an atom that had already been measured along a slightly different axis. But that information is not coming from the future — information never comes from the future! — it's a statistical consequence of events that have *already happened.* Whether it's 1,000 previous flips of a trick coin, or having previously measured an atom along another axis, that information has already appeared in the world. And, now that it's a feature of the world, it *informs* us, to some degree anyway, of what might happen next. The information is coming at us from the past, as information always does, and it's constraining our expectations of future events, which would otherwise be uncertain.

Think back to the *Reign of the Mantelopes* example (pages 61–62): If you ask the *Mantelopes* wiki whether a particular mantelope's mother is human, the answer might go either way; it might be completely uncertain. But if the wiki has already decided that the character's father is human — if that information has appeared in the world of the wiki — then the answer to the question about the mother will be constrained all the way to certainty. We know that the wiki will answer no, the character's mother is not a human. She is an antelope. To answer otherwise would be illogical; given that the character is a mantelope, one parent must be human, and the other must be an antelope.

Are there any future events where we're certain of the outcome? Can information reduce the uncertainty of a future event all the way to zero? We already saw (as illustrated in **Fig. 2**, at left) that the outcome of a spin measurement will be certain if it is repeated exactly, at least if the experiment is done under ideal conditions. Also, in experiments with entangled particles (as in **Fig. 18** on page 58), the outcome of a spin measurement is certain, provided the entangled particle's twin has been measured. But such certainty of yet-to-happen events is limited to highly controlled situations, such as these quantum experiments. In everyday life, things are messy and nothing is certain. Even in the iconic example of knowing the future — the prediction that the Sun will rise tomorrow (think of *Annie*) — we know that it's *very likely* to happen, but it isn't *certain*. There is a tiny but non-zero chance that some unforeseen astronomical event could disrupt the Earth's rotation or obscure all light coming from the Sun, so that it wouldn't "come up" the next day. In fact, the near-certainty that we feel about the Sun rising tomorrow is the result of in-formation from the past, specifically, the Sun's excellent track record of reliably rising every day. All of that information, a sunrise roughly every 24 hours (polar regions excepted) for all of recorded human history and presumably beyond, is what reduces or constrains our uncertainty about the Sun's behavior tomorrow. It's *virtually* certain, but it's not 100% certain.

Now let's think back to the double-slit experiment, specif-ically the version where one photon at a time goes through the apparatus (page 39). An individual photon could wind up anywhere on the screen. Before the photon impacts there, what information do you have? Is your uncertainty reduced at all? Well, if you know all of the variables involved — the distance between the slits, the distance to the screen, and the wavelength of the light — and you know the relevent equation, then you can calculate a probability distribution at the screen (page 41). That will tell you there are some places where the photon is very unlikely to show up, and other places where it's more likely. The information that you have about the experiment and its setup has partially constrained your statistical expecta-tions of the result of any individual photon impact. Then, when the photon finally does go *splat*, those expectations are constrained much further. This is what physicists mean when they refer to the "collapse of the wave function": It appears that a photon's spread-out probability distribution, which is the most we can say about it before we measure its final position, suddenly and completely collapses to a single point, providing highly specific information when it actually shows up on the screen, at which time our uncertainty of the photon's final resting place has been reduced to nothing. Note that you can still only describe this position in a relative sense, for example, four millimeters to the right of the center line.

Different theories try to explain where the collapse comes from and if anything causes it. An informational interpretation might compare the before-and-after pictures and say, partial information (i.e., where the photon *might* go splat) has been updated with the appearance in the world of more information (i.e., where it *did* go splat). This is similar to the case of flipping a trick coin: By performing one flip and recording the results, partial information about how the coin *might* land "collapses" into information about how it *did* land. If you prefer, you can say that partial uncertainty about the outcome, represented by an intermediate percentage, has collapsed to zero uncertainty, represented by either 100% heads *or* 100% tails.

The Uncertainty of Measured Values

In Chapter 3 we talked about information being relative, with examples of measurements whose outcomes are always relative to the observer performing them. For instance, if you measure the velocity of a thing, that does not mean discovering its absolute velocity value, but rather, finding a value that represents a velocity against (or with respect to) your particular frame of reference. You will recall that this extends to many other measured values, including mass and energy. Now I'd like to connect that concept with our definition of information as the reduction of uncertainty. What does uncertainty reduction mean in the context of performing a numerical measurement?

With some measurements, we've seen that the outcome provides information that reduces our uncertainty all the way to zero. This is the case with looking at a coin that's just been flipped, or measuring whether an atom's spin is up or down.

There's no ambiguity in these measurements; either a coin landed heads-up or tails-up, and an atom will measure either spin-up or spin-down, with no in between. The result of such either–or measurements produces only one bit of information, such as a 1 or a "yes" (which might represent heads or spin-up) vs. a 0 or "no" (tails or spin-down). In these simple situations, one bit of information reduces our uncertainty to nothingness. Still, it's important to realize that the bit is not absolute information, for the same reason that saying "yes" or "no" has no meaning unless it is in reference to a specific question. Put another way, getting a measurement value of 1 may mean heads in one context (where 1 = heads) but tails in a different context (where 0 = heads). Likewise, in measuring an atom's spin, 1 could represent spin-up, or it could represent spin-down, or any other direction, depending upon the reference frame. As with all measurements, the information must be considered with respect to some reference information in order to have any useful or communicable meaning at all.

Those are the simple cases — the binary, either–or ones. But most measurements that humans make aren't so simple. If you want to measure the distance around the circumference of a circle, you usually can't express the outcome with a single bit. The result can be any number, perhaps with decimal places. Suppose we determine that the circumference is 3 units of length. In the real world, that value probably isn't an exact, perfectly precise measurement; the actual distance may in fact be slightly more or less than 3 units. Still, the number 3 gets us close; it reduces our uncertainty somewhat. We may not know the exact length, but we *do* know that it isn't closer to 2 units, or 4, for example. So, what do we do when "3" isn't enough uncertainty reduction? We make a more precise measurement. Perhaps we find that 3.1 units is a closer approximation.

Intuitively, you can see the link between the amount of uncertainty reduction and the quantity of information in a typical measurement: "3.1 units" has the capacity to reduce our uncertainty of length considerably more than merely "3 units" does — and, consequently, the number embodies more information. We require more information or bits to express the number 3.1 than we need to express 3: It requires two base-ten digits on this page rather than just one. Taking this to an extreme, a measurement of 3.14159265 provides much greater uncertainty reduction still. That string of digits, considered with respect to a reference system of length units (such as a ruler), provides proportionately more reduction of uncertainty about the circumference, and consequently it requires proportionately more bits to store in a computer file. The more you want to reduce uncertainty in a numerical measurement, the more information you need to collect to get there.

Information can reduce our uncertainty not only about a quantity being measured directly, but also about other related quantities. For example, you don't need a long tape measure to determine the distance to another city; you just need to know how many minutes it takes to get there, and how fast you were going. (For this example, let's pretend odometers don't exist.) The more precise your speedometer and your clock, the more precise your distance measurement will be. There's an equation that expresses how these three numbers relate to one another, and the less uncertain you are about two of the numbers, the less uncertain you will be about the third.

$$S = (55\text{--}65 \text{ mph}) \times (0.7\text{--}1.2 \text{ hr}) = (38.5\text{--}78.0 \text{ mi})$$

$$V = \frac{S}{t} \text{ (or) } S = Vt \qquad S = (58\text{--}62 \text{ mph}) \times (0.8\text{--}1.1 \text{ hr}) = (46.4\text{--}68.2 \text{ mi})$$

$$S = (59\text{--}60 \text{ mph}) \times (0.9\text{--}1.0 \text{ hr}) = (53.1\text{--}60.0 \text{ mi})$$

Fig. 3. An equation (left) expresses the mathematical relation between velocity (v), distance (s), and time (t). If you want to calculate a distance based on measurements of velocity and time (right), then the narrower the range of your measurements, the more precise your calculation. The more uncertainty reduction you put into the equation, the more you get out of it. Another way to say this is: The more information you put in, the more information you will get out.

Similarly, you can measure the wavelength of light indirectly by using a double slit where the distances in the setup of the experiment are known quantities. Suppose you want to know the wavelength, but at the moment it's completely uncertain. So, you fire up the laser. Each individual photon that registers on the screen tells you very little — but if you let a great many photons go *splat*, you'll be able to see interference fringes clearly enough to measure the distance between them, at least within a range of uncertainty (those fringes can be hard to measure). Plugging that partially uncertain measurement into the appropriate equation, along with the other distances, will give you a correspondingly constrained measurement of the wavelength. You might say that each photon has reduced your uncertainty of the wavelength by a tiny amount, and the more photons you register, the more uncertainty reduction you achieve. Just as each grain of salt in a recipe adds just a little flavor to the whole, with more and more photons comes more and more information about the number you're seeking.

These concepts can be tricky, so let's review:

1. Information is something that reduces uncertainty. Anytime there is a situation that's uncertain, and some number or quantity comes along that reduces that uncertainty, the number or quantity that came along is information.

2. Information can reduce uncertainty anywhere from slightly (like the odds that a trick coin will turn up as heads), to reducing uncertainty completely (like the result of a single coin flip).

3. Information is never absolute. In order to reduce uncertainty, information must be considered against a reference frame, i.e., with respect to some specified coordinate system. Without this, a quantity of information is as meaningless as a standalone answer "yes": Yes to what question?

4. All but the simplest measurements can be made to any arbitrary level of precision. More precise measurements yield more information, i.e., more uncertainty reduction.

A Closer Look at Informational Descriptions

We are starting to see how any individual piece of information, by the very definition of information, represents some relation between two systems of some kind. By that I mean, if you have information on the velocity of an object, then that information is *specific* to the relationship of motion between the object and yourself. After all, some other observer may measure the same object's velocity differently, in which case, the information about velocity will not be the same. But, we also have to remember that even information representing that relationship is not absolute in any sense: It is also relative to your chosen reference frame. It's not enough to say, "The velocity, relative to me, is 7.5" — 7.5 what? Not only have we not specified the units of velocity (it could be miles per hour, miles per second, etc.), but we also haven't specified the direction, which is a component of velocity. That is why we need to specify a reference frame, so we can say something that actually reduces uncertainty, such as: "The object's velocity, relative to me, is 7.5 *feet per second*, along *the x axis*," with the axes oriented in such-and-such a manner.

A basic measurement such as velocity or length doesn't represent just one relation; it can be broken down into multiple informational relations, all compounded upon each other. Consider the proposition of measuring the distance between the goal lines on a football field, in yards, along the length of the field. There is a spatial relation between the two goal lines; they are two different things in the world, separated via an extension in space (see page 81), which we call distance. That separation is represented by one relation. *That* relation, in turn, can have its own relation with the separation of the two ends of a yardstick, which is its own relation. There's a diagram coming up, so try to stay with me. And, what's the relation between the goal-lines relation and the yardstick-ends relation? It's a ratio: One has a considerably larger magnitude than the other. In fact, it's 100 times larger in this case (100 yards). So this last relation can be meaningfully represented by the integer 100, as long as we don't forget to tie it into the other relevant relations. One of these relations specifies the units — we've chosen the separation between the ends of a standard three-foot yardstick to represent one unit in our measurement scheme, an arbitrary choice (as in **Fig. 5** on page 89). Is that all? Not really — some conscious human being is thinking about this, and there is a relation between that person and the statement: "The separation between the goal lines equals 100 yardstick-end-separations." It doesn't even have to stop there; the person might communicate that information to another person.

This compounding of informational relations can be diagrammed schematically, where each system (or thing) is shown with a dot, and the relations between them are lines connecting the dots. In this manner, we can sketch out how relations can build into a tree-like structure of ever-increasing complexity:

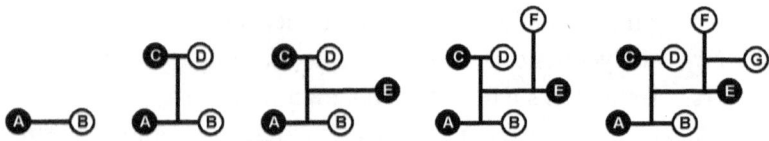

Fig. 4. A schematic illustration of informational relations building upon informational relations. **A** and **B** are the goal lines on a football field, which have a certain relation in space. That **A–B** relation has a relation with the ends of a yardstick, represented by **C** and **D**. **E** is the ratio inherent in that relation, the number 100. **F** is an intelligent being contemplating this number of yards between the goal lines, and **G** is a friend who heard him say that the distance between the goal lines is 100 yards.

Informational relations compounding upon other informational relations is a key to the simplest-case scenario. **Figure 4** above starts to hint at the way the world is fundamentally put together. What the diagram simplifies is that **F** and **G** — human beings — are each highly complex systems of relations themselves, involving senses, memories, histories, instincts, and thoughts (to say nothing of their physical bodies). This starts to get at the problem of explaining the nature of consciousness, which will be discussed in the last chapter. For now, at least, you can appreciate that the diagram captures how information can build upon other information, to create what we language-using sentient beings call *context* and *meaning*. The number 100 (represented by **E**) in isolation is virtually meaningless, but as a part of this tree of connected informational relations, it means something quite specific: "The goal lines **A** and **B** are 100 yards apart, as considered by **F** and communicated to **G**."

At this point, you may be wondering, is the relation between **A** and **B** in the diagram — two inanimate lines on a field — really *informational*? In what sense does that relation meet our definition of information as something that reduces uncertainty? Well, goal lines painted on a field do not lie awake at night fretting about their uncertainty of things. That's something that conscious humans, and probably only conscious humans, do. However, the mainstream scientific picture of *stuff* describes even simple inanimate objects as having informational rela-tions. Two electrons, for example, are said to have information on each other. It's a way of saying that if they are close together, they seem to know each other's presence, in the sense that they electrostatically repel one another.[1] In the example of the goal lines, the mere fact that they are separate, independent features of the world means that they necessarily have *some* relation(s) to each other. For example, they have a separation in terms of space (and therefore also in terms of time, as discussed on pages 80–83). In fact, while they may not "know" each other's presence as dramatically as two repelling electrons, there is a very simple way to prove that the goal lines have informational relations: Measure them. You will find that you can extract all kinds of information about their relationship. For example, there is the ratio of the lines' lengths. We would

---

1 The simplest-case scenario's informational view is that the pair of electrons *literally* are composed of informational relations — at least to the extent that they have been measured, up to the limits of the uncertainty principle (pages 44–46), or have manifested in the world in other types of interactions where information has been registered somehow.

express this as a numerical ratio, which in this case would be 1:1 (both lines are equally long). You can also measure the angle between them, determining their relative orientations in space. This would also produce a kind of 1:1 or unity ratio, except that in terms of angles, we call this zero degrees (or radians) by convention. Having measured that the goal lines are parallel, you can measure the distance between them, and probably find something very close to 100 yardstick-lengths. How close? Well, if you want to maximize your uncertainty-reduction regarding that question, you will have to make an extremely precise measurement, resulting in a number with many decimal places and containing a lot of information.

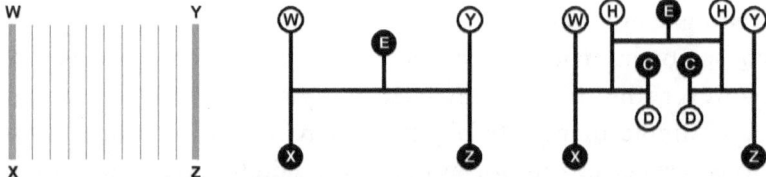

Fig. 5. **WX** and **YZ** are the ends of a football field's two goal lines (left). Those two relations of separation, in turn, have a relation (**E**, center), which is 1:1, because the lines have equal lengths. If we were making a numerical measurement, we'd compare **WX** and **YZ** to a standard length **CD** (right), such as a yardstick. We'd then find a ratio of 40 (yards), **H**, for the length of each line — and the relation between 40 and 40 is again the 1:1 identity, **E**.

For another example of mutual informational relations, consider two glasses of water. Suppose the only thing we're concerned with is the level of water in each glass, and not their sizes, separation in space, etc. If both glasses have the same level of water, the situation is symmetrical. The relation between the water levels is the identity relation; if you connected the glasses with a tube, no water would flow in either direction. But if one of the glasses has a higher water level than the other, then we have an asymmetry. There is some kind of non-identity relation between the two levels, and that relation could be measured. You could say that the higher water level has information on the lower water level, but it's just as valid to say the reverse. The informational relations between these two things are equal but opposite; they are mutually inverse. Does that remind you of anything? There is Newton's Third Law of Motion (that every action has an equal and opposite reaction), or the opposite spins of entangled particles, or the tradeoffs described on pages 84–85. All of these things are connected — and they all go back to information as the basis of everything in the world.

Which brings me to a point: This analysis of information may seem overly complicated, and the point of the exercise unclear. But, just as the science of stuff breaks down matter and energy into fundamental particles to better understand how the world is put together, we need to do the same with information. And, just as chemistry builds atoms and molecules out of the fundamental particles of stuff, we will do something similar with information, beginning in the next chapter.

Debts As Informational Relations

We've used goal lines and glasses of water as analogies to understand how relations work. Another analogy is to think of debts between friends. If Alice owes Bob 100 dollars, then equally true is the inverse — the "equal-and-opposite reaction" statement that Bob owes Alice *negative* 100 dollars. You can imagine a network of debt-relations among four friends, where each friend owes money to, or is owed money by, each of the other three:

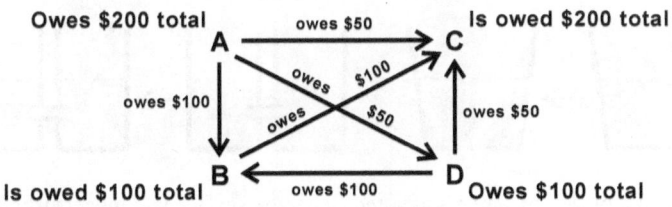

Fig. 6. The financial relationships among four friends Alice (**A**), Bob (**B**), Charlie (**C**), and Donna (**D**). The total debt among them equals zero.

An interesting aspect of this relational network is that if you add up all of the dollar values, you will get zero. That is indeed what we should expect, given that when the friends first met, presumably they were debt-free with respect to each other. There's a big tie-in with the universe here: The friends' debts in this example are analogous to various measurable properties of objects in the world, such as electric charge or intrinsic spin (page 53). Let me explain.

The case of the friends is an example of a *closed system*. We are only considering their internal debt relations to each other, and not, for example, whether Bob owes anything to the IRS. For simplicity's sake, let's say all four of them met at the same time, perhaps while waiting in line for a concert. If we assume that they started off financially even with each other, then the debts must have arisen later on, when they needed money from each other or what have you. But, since the system is closed, the internal sum of the monetary relations will always equal the same number. In fact, if the friends continued to trade money among them, then at *all* times, at least within this closed system, the total debt would *always* equal zero. We might say we started with a "flat" (debt-free) system, and then we perturbed that system, the way the surface of a bucket of water can be perturbed by agitating it. No matter how you jostle a bucket, though — provided the system remains closed, and no water flies out or is added — the average height of the water will always be identical to the height when the surface was flat.

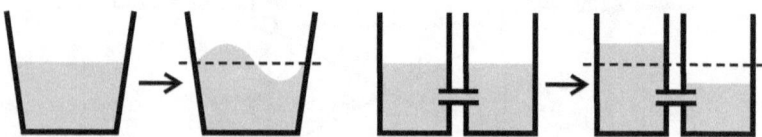

Fig. 7. If a bucket of water is agitated (left), the surface will be perturbed, but the average level is unchanged. This is similar to the case of two glasses of water connected by a tube (right). In a closed system, even if imbalances appear, the totals will always remain the same, i.e., they will be invariant.

Many cosmologists believe that the universe began as a quantum fluctuation that's similar to the kind of perturbation discussed above. Due to conservation laws, the totals of related quantities should remain the same as when the universe began. For example, if we could add up the gravitational potential energy of every object in the universe, and add it to the other kinds of energy, then the numbers would likely cancel to exactly zero. This makes sense because presumably, at the Big Bang, the universe started with zero total energy, like the debts of our friends above. Also, adding up the total positive and negative electric charge should result in an exact cancellation; the universe presumably began with zero net charge, and the charges that exist now are equal-and-opposite perturbations from that "flat" situation. This is pretty much the way it has to be if you accept that the universe is a closed system with no outside. As the physicist Lawrence Krauss discussed in his popular book *A Universe From Nothing*, you *can* get something out of nothingness — but everything must have an equal-and-opposite counterpart, within that closed system.

Can Informational Relations Be Real?

Now let's indulge ourselves with a philosophical question: Is the network of debt relations among the four friends *real*? I don't mean "real" in the sense of the way people might discuss a ghost or a UFO. I mean "real" in the sense of Chapter 1 (page 27): an internal logical consistency such that the relations remain the same, even despite a uniform transformation — indicating that the reality equivalence principle holds. The relational network of debts among the friends is real by that definition: If the dollar values were converted to Italian lira, for example, then all of the numbers would scale up consistently, by a factor of around 1,700. The indebtedness of any one friend, relative to any other friend, would be exactly the same as before. Alice would still owe Charlie half as much as she owes Bob. Informational relations within closed systems are always logically consistent like this, and that makes them real by our Chapter 1 definition. Any changes in those relations (i.e., exchanges of money) will have consistent, predictable results, as if the debts were following laws of nature, and these "laws" apply to the entire system, for any length of time. Meanwhile, no matter what currency unit we convert the friends' debts to, the total will still always be zero. Try it!

Consider another example with four point-objects, whatever those might be, arranged on a plane of Cartesian coordinates (**Fig. 8**). Information about their distance relations, expressed as a ratio (arbitrarily) with the pair **AB**, is shown between each pair. Also shown is information about the angles they form; each angle is expressed as a fraction of a right angle, as defined by the grid. For example, 45 degrees relative to the grid = 0.5. At upper-right, the objects and their relations have undergone a stretching transformation — but, in terms of the numerical ratios/fractions, nothing has changed. If you were an inhabitant in that world, you would not be able to distinguish between the upper two configurations; both would appear equally real. Finally, we consider a world *without* the point-objects, containing only the informational relations (bottom). Naturally, transforming the relations would still have the same null effect on the numbers. In fact, an inhabitant that could only sense information could not distinguish between *any* of the four configurations; the scenarios would appear equally real, including the ones where stuff-like objects are not fundamental features of the world.

I've used the stretching analogy several times in this book, and it may appear to be a geometrical trick without any practical relevance. But actually, the transformation pictured in **Fig. 8** is another way to diagram what happens in special relativity. Observers have their own coordinate system of space and time, within which they observe and measure themselves and their surroundings to be at rest. However, when an object is moving relative to this frame, then the coordinate system is essentially deformed, resulting in the observed phenomena of time dilation and Lorentz contraction (see **Fig. 9**).

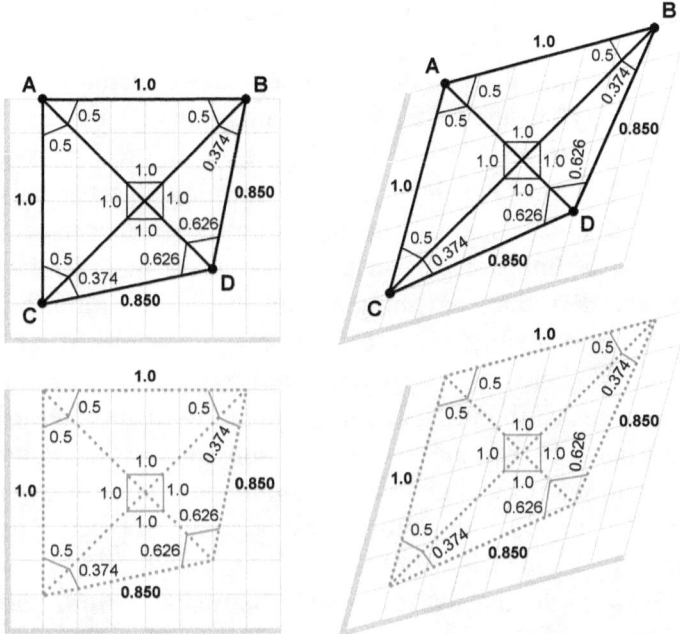

Fig. 8. Uniform transformation of a set of relations between objects (top), and the transformation of the same abstract relations without objects (bottom). The idea of angles expressed as unitless fractions, where 1.0 = 90 degrees, comes from Norman Wildberger's rational trigonometry.

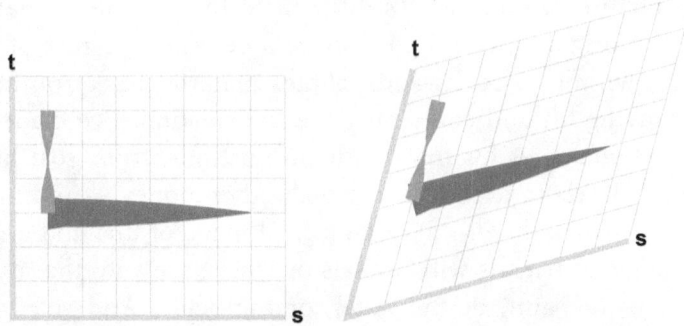

Fig. 9. A rocketship and hourglass at rest, i.e., in the rest frame (left). When they are observed moving at relativistic speeds (right), the coordinate system used to measure them becomes "stretched." Relative to that grid, the ship appears shorter, with the rate of the hourglass slowed by the same factor (in the t direction for time). At the speed of light, the grid would be infinitely stretched in the diagonal direction, so the ship would be measured as infinitely short relative to the grid, and its hourglass stopped.

Still, not many people would consider information to be a "real thing," in the same way that an apple or a baseball is a real thing that you can touch and pick up, and feel its weight. It's intuitive to believe that such solid-seeming objects persist in the world, independently and across time. We seem to require that kind of intuitive persistence-in-the-world in order to buy the proposition that something possibly might be a fundamental constituent of reality. It's so easy to imagine an electron being around long before you were born, and continuing to exist long after you die, all the while carrying with it information regarding its rest mass, charge, and so on. But, *informational relations* existing and enduring like that? Without some kind of external hard drive on which the information is stored, or piece of paper where it's printed? Come on!

How could the world possibly be made of information, rather than stuff? It's so vague and ethereal. The stuff picture, with atoms colliding and combining, seems to be a picture of *something*; we can imagine it going on, even if we have to concede that the quantum world of these particles is weird. The alternative picture, of informational relations, seems to be a picture of nothing. It's difficult to imagine how a naked bit of information could possibly be a persistent feature of the world, without a rocks-and-pebbles material background of stuff giving off and containing the information. I've made all sorts of analogies for informational relations, from goal lines to water levels to friends owing each other money — and in all of those cases, there is a football field made of dirt and grass, or glasses or friends with wallets or what have you, that carry the lines of paint, or the water, or the debts. And yet, John Wheeler claimed that perhaps everything in the universe exists *only* as information. He saw the world as being like a collection of coin-flip results without the coins, or a network of debts among friends, but without the friends. This is one of the most abstract, unconventional, tough-to-grasp aspects of Wheeler's vision, and of the simplest-case scenario. Several points.

1. Since you've been immersed in the stuff picture for your entire life, you're accustomed to thinking of every event being a function of matter and energy, with every event caused by the influence of other matter/energy. But remember that matter is over 99.999% space, and according to quantum mechanics, we can only describe the location/momentum, spin direction, etc., of an elementary particle as probabilities until they are specifically measured or have some kind of interaction. And even then, according to the theoretical framework known as quantum field theory, particles of matter and energy are described fundamentally as excitations of various fields, and are not really pellets of stuff with an inside and an outside (whatever substance their interiors might be made of). If anything, the everyday interactions you have in the world are more about experiencing forces, and informationally registering apparent energy particles we call photons, than having direct interactions with matter — whatever *that* might possibly mean. Ultimately, information is all we have to work with. Perhaps every informational relation just *is*, the same way you assume without issue that every atom or other particle just *is*.

2. There are things in the world that aren't material. The circle is an example. Also, the natural numbers: 1, 2, 3, and so on. One can argue that ideal shapes and numbers are invented, but stars and planets form near-perfect three-dimensional circles (spheres), and in every atom of lithium that you could ever examine, you will find that it contains exactly three protons. Despite being immaterial, numbers just *are*; they're features of the world.

3. There are immaterial features of the natural world that aren't rooted in matter at all. For example, the physical law known as the conservation of momentum persists everywhere, and timelessly. When one bowling ball hits another, there's always a recoil, and the total momentum before is guaranteed to be the same afterward. It's no wonder Newton believed that the laws of nature arose from a perfect god pulling the strings of the universe.

When you think about it, these laws are like your dead relatives watching over you — they're a little creepy! Physical laws aren't written on a stone tablet somewhere, yet we have no problem thinking of them as just *being in the world*, and affecting everything that happens, everywhere.

4. There are even informational relations, also not rooted in matter, familiar to you. The irrational number π is an informational relation for anyone who uses it: If you know a circle's diameter, π is information that reduces your uncertainty of the circumference. For that matter, any mathematical equation[2] is information with respect to its user: If you know the rest mass of something and you know the speed of light, then the equation $E = mc^2$ reduces your uncertainty of the energy equivalent. Despite not being written in the stars, mass–energy equivalence is a feature of the world, everywhere and at all times. It just *is*.

5. There are immaterial things that persist everywhere timelessly, not only in our universe but perhaps all possible universes. For example, the identity relation that $A = A$. This is the domain of formal logic. It's also the case that a false statement cannot be true. I apologize for getting philosophical here, but logic is one of the backbones of Western thought, science, and math, so you should not be embarrassed by it or afraid of it.

Finally, a note from theoretical physics: An outgrowth of string theory, called the *holographic principle*, speculates that all of the information describing our universe is "painted" on a two-dimensional surface at some cosmic boundary. In this conjecture, the three visible dimensions of reality project into our world similarly to the way a 3D image emerges out of a hologram on film. If you need to imagine that the world's information is written down or physically extant somewhere, this is one thing to consider.

---

2 The cosmologist Max Tegmark has hypothesized that the universe is fundamentally mathematical in nature. This is a decidedly non-stuff view, and not far at all from the informational picture. If I may brag, Tegmark is on record on YouTube saying that an invention of mine, which demonstrates spacetime curvature, should be in a science museum. (I only feel comfortable dropping names like that in footnotes.)

So much of the world is considered immaterial, even in a conventional sense — from the vast space inside atoms, to physical laws which persist throughout space and time, to the immutable laws of logic. I realize it's difficult, but try to imagine something like a network of numbers and ratios of numbers that just *are*, in the manner that the laws of physics and logic just *are*. It shouldn't then be too much of a stretch to envision a universe which, in its entirety, is a complex web of such relations. John Wheeler would be probably be pleased with that picture.

Information for Non-Humans

Let's not forget that nearly 100% of life that has lived on Earth has never flipped a coin or written down a number. So, I'll close this chapter by discussing what information is to them. Humans invented the term "information," and we invented numerical measurement and all of the measurement units we use to convey information. We have phones and cable TV, while animals, plants, and bacteria don't. But even ignoring the fact that many animals communicate — from the songs of humpback whales to chemical signals used by insects — information is critical to the survival of animals and the rest of life as well.

To use a domestic example, when my cats walk into the house, they're extremely quick to notice if anything is different. It appears that they have some kind of map of a recent state of my house in their brains, and when they come in, they're making a comparison between that map and the new information coming to them via their senses. When there is a difference, they observe this difference. If there is a bucket on the floor that wasn't there before, they are uncertain about what the object is, and they investigate via sight and smell and touch, so as to reduce their uncertainty. This is a key part of animal adaptation and reproduction: Animals need to observe changes in their environment, changes that might kill them, provide a potential mate, provide potential food, etc. The observation of these changes results in information for that animal. Animals are also able to make crude measurements: If a rabbit observes a fox, and then it observes that the fox is closer, it's in the rabbit's best interests to compare these observations, assess the fox's rate of approach, and perhaps get the hell out of there. As another example, the famous "waggle dance" of honeybees, discovered by the biologist Karl Von Frisch, conveys to hive-mates information not only about the position of newly discovered flowers, relative to the position of the Sun, but also a rough measurement of how far away the flowers are: The closer the flowers, the more frantic the bee's movements, and the shorter the waggles last. Fellow worker bees, who would otherwise be quite uncertain about where flowers might be, suddenly find their uncertainty reduced by the waggle dance of one bee that discovered them.

Plants use information, too? Definitely. Consider a plant's tendency to grow toward light, called phototropism. If you shine sunlight on only one side of a plant, it will observe this, and over time it will preferentially grow in that direction. This information depends on some kind of asymmetry or spatial distinction, or an asymmetry over time (change), in the environment. If there's no asymmetry — the house looks the same as the last ten times the cats came in, or the plant is illuminated equally from all angles — then no distinction can be made in the environment, and so there is no impetus for the organism to alter its behavior.

Does anything other than biological life use information? Another category is technology. When you press a button on a desk calculator, you set up an electronic potential, a voltage, that wasn't there before. The device is manufactured to measure this change, and it alters its behavior in response. But aside from technological devices, which are relatively complex, it would be difficult to argue that things in the non-living world employ information in the strict sense. Tilting a platform may cause a ball to change its behavior and start to roll downhill, but both before and after the tilt, the ball is merely responding passively to the configuration of the environment. Unlike a cat sensing something different, or a plant sensing and physiologically registering a directionality of lighting, the ball merely wants to fall, as it always does, and tilting the platform passively provides a way for it to take a new path through spacetime, the same way that an inflated balloon deflates when you let it go. Although they're obviously subject to the same laws of physics, living systems and some technological systems respond to their environment in complex ways, involving memory storage, hormone production, etc., which are active processes that consume energy. We'll look closer at these distinctions in Chapters 8 and 9.

Things to Remember From Chapter 4

• Information is something that constrains statistical expectations or reduces uncertainty in some way.

• Information can reduce uncertainty partially, or it can eliminate uncertainty altogether.

• More precise measurements require more information. More information must be obtained to perform the measurement, and more must be written down to express the result.

• Informational relations (Chapter 3) can be compounded upon each other or compared. You can compare two measured lengths, each of those lengths in turn being numerical relations (ratios) with the ruler-length that you chose as your reference unit.

• A closed system of informational relations is "real" in the sense of our definition from page 27: It obeys logical consistency. If one relation changes, other relations change in order to exactly balance the system. An inhabitant of that closed system would view it as a "real world."

CHAPTER 5

FROM QUANTUM MECHANICS TO INFORMATIONAL MECHANICS

Quantum mechanics is described as one of the most successful theories mankind has ever devised. By "successful," that means the mathematical equations of QM make numerical predictions about tiny objects' behavior that are incredibly accurate and reliable. Plug in some numbers, and the equations allow you to compute not only probability distributions, but also the manner in which those probabilities change over time. Quantum mechanics doesn't allow us to determine exactly where or when a photon or an electron will show up, as the equations describing the classical mechanics of everyday objects can — although, given a region, we can calculate the exact probability that the particle will be found within that region at any point in time. Regardless, no one can say why the equations work so well, in the sense that we aren't sure exactly what's happening at the fundamental level that results in the predictions turning out as well as they do. The formal mathematics, the tools we use in QM, are like a box that we can put raw ingredients into, and the box bakes a perfect loaf of bread every time, even though we have no idea what's going on inside.

For almost as long as there's been quantum theory, there have been interpretations of quantum theory. An interpretation is some attempt to describe the exact physical mechanisms at the fundamental level, or at least it's a statement regarding the most we can say about what's really going on. These interpretations have been coming for 80-plus years, and with few exceptions, there presently appears to be no definitive way to test whether an interpretation is right or wrong. It's one of the most intriguing aspects of quantum physics, as well as one of the most maddening. In this chapter we'll look briefly at a few major interpretations, focusing on a recent one, relational quantum mechanics. We'll examine how information plays a role in that interpretation, and how it connects with the experiments from Chapter 2. Finally, we'll attempt to adapt relational quantum mechanics in purely informational terms — to produce an interpretation that's information-based rather than stuff-based — in order to lay the foundation for what might be a unified theory, *informational mechanics*. According to the simplest-case scenario, informational mechanics describes not only how the world operates on a fundamental level, but also how it operates on *all* levels, and how the universe came to be the way it is today — as well as why the distant past looks the way it does.

Interpreting Quantum Weirdness

Two of quantum theory's founders, Niels Bohr and Werner Heisenberg, attempted to interpret quantum mechanics. Since they were Danish, their ideas were collected into what we call today the *Copenhagen interpretation*. However, examining the history, there never was a unified Copenhagen interpretation. Each physicist's view evolved over several decades, and others, such as John von Neumann and Eugene Wigner, jumped onboard and offered their own modifications of the interpretation. Von Neumann and Wigner, in particular, pursued quasi-mystical views in which the observer's consciousness plays a role in a quantum measurement, perhaps even causing what came to be known as the collapse of the wave function (see page 96) — the apparent sudden transition of a physical system from an indefinite or probability state to a definite, localized state. These more subjective treatments created a backlash starting in the 1950s, when many physicists aimed to rid QM of subjectivism. This indirectly led

to so-called hidden-variable interpretations, which roughly describe reality as the book-world of Chapter 2. Although these
interpretations still have adherents, they are slowly fading from
view, as sophisticated experiments and mathematical theorems
find them increasingly unsupportable.

Very broadly, though, the original Copenhagen interpretation emphasizes the probability aspect of quantum outcomes,
the symbolic (rather than the literal or pictorial) nature of
mathematical tools such as wave functions, and generally, the
inappropriateness of ascribing pre-existing properties to objects
that have not actually yet been measured. In that respect,
Copenhagen is more in line with Chapter 2's wiki-world picture
than the book-world picture.

Famously, Albert Einstein was not a big fan of quantum
mechanics. In reaction to the probability aspect, he told Bohr, "I
am convinced that God does not play at dice." (Bohr responded,
"Don't tell God what to do!") Responding to the claim that
properties can be said to exist only when an object is observed,
Einstein supposedly snarked, "I like to think that the Moon
is there even when I am not looking at it." His challenge to the
concept of entanglement (discussed in Chapter 2) was his greatest
contribution to quantum theory. The so-called EPR paradox,
which is named for Einstein and two other authors (Podolsky
and Rosen), asks how entangled particles can seemingly
communicate instantaneously via "spooky action at a distance."
Even today, this question forces physicists to confront questions
about the ultimate nature of space and time itself.

The next major interpretation to come along is probably the
one most familiar in the public mind. Although he didn't use this
term, Hugh Everett introduced the *many-worlds interpretation*
(MWI) in his 1957 dissertation at Princeton, where his doctoral
advisor was none other than John Wheeler. Today, thanks to
science-fiction and Hollywood, it's commonly known by names
like "the parallel-universes theory." MWI claims that anytime
there is an observation (broadly defined), or anytime something

occurs that could have multiple outcomes, the universe splits. All of the possible outcomes are realized, in their own separate universes, and the single outcome that we observe is just one branch. We split, too, and each of our "offspring" experiences each outcome in a separate universe. If I observe an electron over here, there's another universe in which I observe an electron over there. It's a mind-blowing proposition for sure, one that has captured people's imaginations, as they picture a parallel-universe version of themselves having made a different decision in the past, leading to a completely different life, and so on. However, the many-worlds interpretation — which wasn't on many physicists' radar until Bryce DeWitt revived it years later — has come up against resistance in the physics community. One common criticism invokes Occam's razor (see pages 23–24): Why have an infinite or near-infinite number of ontologically real universes, multiplying like well-fed bacteria, when perhaps under another interpretation, only one universe is necessary? After all, we observe only one universe. Critics have described MWI as extravagant, profligate, even absurd. But MWI adherents claim that Occam's razor is actually in their favor. In Chapter 6, I'll argue that the many-worlds crowd *and* their critics are both correct: The extravagancy isn't in the splitting of the universes, or how many different universes might result. The extravagancy of MWI, as people ordinarily conceive it, is that *one universe made of stuff is extravagant enough!* Of course it's absurd to think of a vast universe of stuff splitting into trillions of vast universes also made of stuff. But if you think of alternate universes merely as all of the *potential* configurations in which information could possibly appear — like all of the ways that a computer-generated *Reign of the Mantelopes* wiki could possibly develop and grow in complexity — then splitting universes is a feature, not a bug.

Another popular interpretation (if you can call it that) is *instrumentalism*, often more rudely called "shut up and calculate." This approach rejects the asking of questions about what's going on inside the black box of quantum mechanics. Instead, the results of calculations, and finding new mathematical tools to get even better results, are all that matters. It's a favorite of skeptics who have seen quantum interpretations get hijacked by pseudoscience (don't get them, or me, started about *The Secret*). Unfortunately, many of these skeptics haven't been exposed to the subtler aspects of natural philosophy, such as those discussed in Chapter 1. Still, instrumentalism puts information first and doesn't concern itself with stuff that might be underlying the information, so it's all right in my book.

A Breakthrough: Relational Quantum Mechanics
In 1996, the physicist Carlo Rovelli introduced a fresh, bold interpretation of quantum mechanics. It wasn't entirely his idea; scientists such as Simon Kochen and Lee Smolin had explored similar themes in published papers. But Rovelli's paper, which appeared in *The International Journal of Theoretical Physics*, was the first time the concept had been assembled in one place with such clarity and completeness. Reading the paper, one is struck by the simplicity of Rovelli's argument, as well as its breadth: Relational quantum mechanics (RQM) offers an interpretation of quantum phenomena that makes no distinction between technological versus natural systems, living versus dead systems, or large versus small systems. Rovelli arrives at the interpretation inductively, beginning with a few propositions derived from experimental observations. His approach — building an idea from the ground up, questioning common assumptions, and aiming for a combination of simplicity and across-the-board applicability, all expressed with straightforward clarity — became a model that I tried to follow in writing this book.

Rovelli begins by recalling the foundation-shaking insights of special relativity. Before Einstein published his famous 1905 paper, it was known that the speed of light is constant, and it was assumed to be constant relative to a "luminiferous aether," a medium through which light waves propagate, the way ocean waves propagate in water. But the Michelson–Morley experiment of 1887, which was designed to detect the aether, found nothing. The experiment showed that the speed of light is measured to be the same at different times of the year, when the Earth should be traveling at different velocities through the aether. If the speed of light were constant relative to this universal grid, we should measure a change when performing the experiment months apart, and Michelson and Morley did not. As Rovelli points out, this motivated Hendrik Lorentz and others to *interpret* the experimental findings, in order to wrestle them into alignment with the assumed truths about the world (e.g., that the absolute background of the aether must exist). For Lorentz, that meant perhaps matter literally, *physically* contracts in the direction that it's moving through the aether, due to some unknown interaction between matter and aether. If Michelson and Morley's apparatus physically contracted like that, and if every device that measures the speed of light contracted like that depending upon its direction of motion, then we would always measure the speed of light to be the same, as judged against measuring devices contracted by the same factor. This would explain why the speed of light is not measured to change as the Earth changes direction —

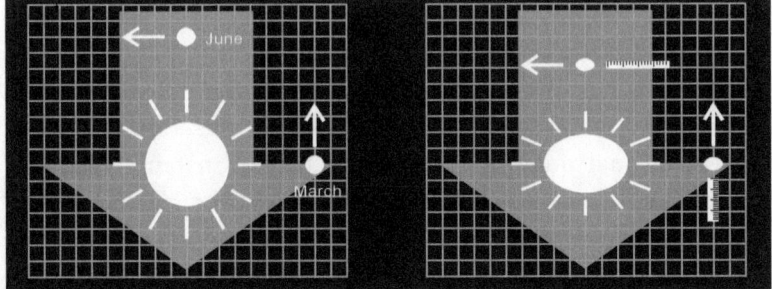

Fig. 1. In the 1800s, it was believed that the Solar System sits in an aether that flows like a river (gray grid and large arrow). If so, the speed of light going in a particular direction on Earth should be measured to change between March and June, as Earth's direction through this flow changes — but it wasn't. Lorentz wrongly interpreted this to mean that all matter, including rulers, must contract in the direction of the aether flow (right).

and, it might never be measured to change, no matter your state of motion. Lorentz created a series of transformations, equations for calculating the "true" distances and durations that occur when objects move through the aether. To this day, the Lorentz transformations bear his name. Rovelli points out that even though Lorentz's mathematical equations work just fine, his interpretation of what's really going on — involving a unidirectional physical contraction of matter — was weird, complicated, and required new and unknown physics.

Rovelli then points out that a remarkably similar situation exists today with quantum mechanics. The equations work incredibly well, but the interpretations of what's really going on are all over the map, and they still generate a ton of controversy and debate, nearly a century after the equations were formulated. Rovelli reminds us that with relativity, Einstein didn't introduce new physics. Instead, he explained the *physical meaning* of the Lorentz transformations, and he did this by attacking a prevailing assumption no one else was questioning: that space and time are universal, absolute, and observer-independent backgrounds, and that time flows by at a constant, absolute rate. Once Einstein vanquished these assumptions, the difficulties melted away. Lorentz contraction became the change in length that's always measured with respect to a reference frame other than the observer's rest frame, i.e., when an object and observer are moving relative to each other. Time dilation, meanwhile, became the frame-dependent equivalent for duration.

Relational quantum mechanics takes Einstein's relativity concepts and tries to generalize them to cover *all* measurable variables that a system may have. Rovelli writes: "Physics is fully relational, not just as far as the notions of rest and motion are concerned, but with respect to all physical quantities." This is basically a one-sentence version of Chapter 3: All physical quantities are relative, and there are no absolutes. Rovelli suggests that the difficulties of QM may derive from an inappropriate assumption: that a physical system can have an *observer-independent state*, or that a system can have observer-independent values of physically measurable quantities. "There is neither an absolute state of the system, nor absolute properties that the system has at a certain time," he writes. This is a huge idea!

Let's think about it for a moment. RQM is a rejection of the book-world view of absolute information possessed by matter, as if there were numbers printed on particles. Instead, in the same way that position, time, and velocity each represent a relation between the object and the observer, other physically measurable quantities such as spin arise out of the relation between the thing being observed and some observer. The easiest way to think of it is: *The state of a system literally is the relation between the system and the observer.*

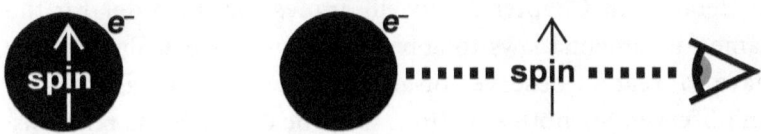

Fig. 2. Relational quantum mechanics rejects the notion that a particle has absolute properties, for example an electron's spin as measured in the up–down direction (left). Instead, that value *is* the relation between the electron and some observer (right), whatever that might be.

Rovelli is careful to mention that the observer in this description can be anything at all, from a person to a desk lamp to an individual particle. Regardless, whatever property is being considered, it must be expressed as a quantity that is *with respect to, relative to,* or *in relation to* that observer. Rovelli's motivation is simple: In quantum experiments, different observers can have different, but not incompatible, descriptions of the same system. Take the Stern–Gerlach experiment (pages 53–57), for example. Before measuring an atom's spin direction, quantum mechanics says we can only describe that property as a probability function. If I measure it to be spin-up relative to some axis, then I have definite information on its spin, at least along that axis, in the sense that my uncertainty has been reduced. My supervisor, however, may have known that I made a measurement, but without knowing the results, then *he* can only describe the spin as a probability; he is uncertain. And, upon asking me, or looking at the apparatus, the probability seems to collapse to a defined number, at least relative to my supervisor.

It may be tempting to dismiss this analysis on epistemic (knowledge) grounds — you might be saying, "the atom *does* have a definite spin after you've measured it, it's just that your supervisor doesn't *know* the value yet." However, this would be an extension of saying before the experiment, "The atom does have a definite spin, *you* just don't know the value yet." The claim that an unmeasured atom has a defined spin along some arbitrary axis is inconsistent with actual experiments, as detailed in Chapter 2. Rovelli argues that in order for the same fundamental laws to apply to all systems equally, then we have to treat an observer-of-an-observer the same as we treat an observer. So, both situations must be described as relations. When I describe a spin, I am not describing an absolute value possessed by the atom, nor is my supervisor; I am describing a value for the *relation* between me and the atom. And, after my supervisor checks the results, he is describing a value for the *relation* between himself and the (me + atom) system. (See **Fig. 3**.) The universe being a place that seems to be logically consistent, both of the measurements will always match up when there is confirmation.

This has connections with Einstein's relativity and even Galileo's relativity. If I measure the velocity of an object, that's an obvious relation; it's an observer-*dependent* measurement. Someone else may measure a different velocity of the same object, yet we can both be right. If I'm on a train playing ping-pong, I might measure the velocity of the ball as 20 miles per hour, while another person on the train platform might measure the same ball to be traveling at 70 miles per hour. These descriptions are not incompatible with each other —

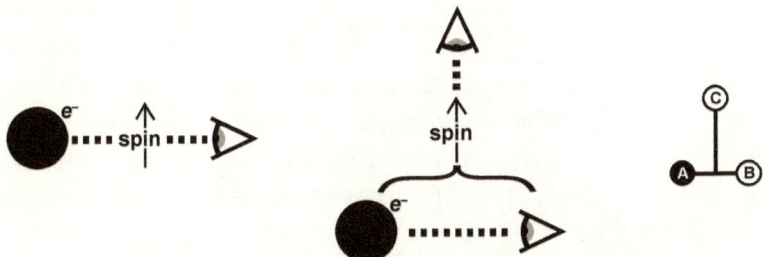

Fig. 3. When an observer measures an electron's spin, a relation is formed (left). A second observer can then form a relation with the electron–observer system (center). The larger system of the relation-with-a-relation can be represented schematically at right, where **B** and **C** are the observers. They are highly complex systems in themselves, simplified in the diagram.

it's just that no absolute velocity values can be pinned down, because they don't exist. Only relational descriptions can be given. The brilliant insight of relational quantum mechanics is that this is how *all* measurable, variable properties in the world work. RQM thus solves what Rovelli calls "the problem of the observer observed": Different observers can have different descriptions of the same system, without those descriptions being inconsistent with each other.

You might ask, if all properties are relational and observer-dependent, why is it only obvious in the case of things like velocity and position? Rovelli points out that actually, even Galileo's relativity principle met resistance. In the 1500s, people still assumed that things are either moving, or they're standing still. In today's world of smooth-riding cars and jets, the relativity of velocity is a plain fact of everyday experience; not so much, though, in the world of carriages and cobblestone roads, where one's "absolute motion" could seem painfully obvious, perhaps literally. And again, if the speed of light were 50 miles per hour, many more properties would be obviously relational, due to relativistic effects. Any measurable quantity incorporating velocity, distance, duration, or mass — as well as things like color, energy, and position in one's visual field — would change just by jogging toward or away from an object. If in that world you saw a friend in danger, you might have to shout something like, "Look out for that red car! ... I mean, blue for you!"[1] We can blame much of our absolute-properties assumption on the speed of light being so high. Regardless, Rovelli's compelling, reasoned argument from first principles has made foundational physics confront the question of whether *any* property in the world, including things like spin and charge, is absolute.

Information at the Heart

After Rovelli finishes his attack on the assumption of absolute, observer-independent properties, he drills down to what's at the core of relational quantum mechanics: information. After properly and rigorously defining what he means by information (more on that in a moment), he makes a few general and largely uncontroversial statements: "Physics is the theory of the relative information that systems have about each other. This information exhausts everything we can say about the world." I cannot imagine a more general way of stating the true mission and scope of physics.[2] Physics is not actually about objects; it's about the *information* we have on them — what we can find out from them, what we can say about them, and what we can predict about their behavior, which can be confirmed only by observation, i.e., more information. Since Rovelli is seeking an interpretation in which quantum mechanics governs the behavior of everything in the world, at all scales, he folds QM into his general statement about physics: "Quantum mechanics is a theory about information — the physical description of physical systems relative to other systems, and this is a complete description of the world." Thus he

---

1 It would probably be too late: Sound would travel much slower in that world, too. Since sound can transmit information, then like anything that can carry information, those waves would not be able to go faster than light (50 miles per hour in this example).

2 Compare Rovelli's quote to this one from the visionary physicist Werner Heisenberg: "Physics must confine itself to the description of the relationship between perceptions." He wrote that in 1927.

defines the role that information plays, at least with regard to our *description* of the world.

Rovelli's definition of information is different from some of the other definitions appearing in the literature. The scientists quoted in Chapter 3 describe information very generally and broadly, as a quantity that reduces uncertainty or constrains statistical expectations for some observer. But the physicist Rovelli seeks an object-oriented definition, describing information in terms of *correlations* between the properties of physical objects. "Any physical system may contain information about another physical system," he writes. "For instance if we have two spin-½ particles that have the same value of the spin in the same direction, we say that one has information about the other one." (An example of a spin-½ particle is an electron.)

Here, Rovelli is drawing from the work of the pioneering information theorist Claude Shannon, as does pretty much every modern scientist who studies information. But Rovelli's object-oriented definition seems to be at odds with his main thesis: He is describing "two spin-½ particles" in a paper asserting that there is no such thing as absolute properties. If it's inappropriate to assume that something has absolute properties, independently of any observer, can we really refer to a "spin-½ particle," or two of them, in an absolute sense? Rovelli's interpretation assumes a *stuff* description of the universe — as in, fundamental particles like electrons being continuous features of the world that possess *at least some* absolute and unchanging properties, such as their spin of ½ and their electric charge of –1. And, for Rovelli, information is something that derives secondarily from the correlation of these things that have these particular absolute properties. Even though Rovelli cites Wheeler's landmark 1989 "it from bit" paper — the same one that I quote again and again in this book — he doesn't adopt Wheeler's conviction that objects like electrons *emerge* from information, the way a computer-monitor image emerges from bits on a hard drive. Whereas information is always relational by nature (as discussed throughout Chapters 3 and 4), particles of stuff with certain absolute properties are not. The stuff picture of the universe seems to be incompatible with RQM's core principle declaring "neither an absolute state, nor absolute properties."

I claim that in order for RQM to be true to its own stated tenets, it must go one step further. It must broaden its rejection of observer-independent states to include not only properties that can take different values (such as spin direction) or can otherwise be measured differently by different observers (such as kinetic energy), but also the values characterizing the *identities* of particles — the information that tells us what something actually *is*. This would be consistent with Wheeler's conjecture that "every particle, every field of force, even the spacetime continuum itself" derives "its very existence entirely" from information. Would Wheeler, who believed that everything in the world derives its existence from bits, agree that electrons or other spin-½ particles are fundamental building blocks of the universe? Clearly not. Wheeler and Rovelli cannot both be right.

At the same time, if we are going to take Wheeler's it-from-bit conjecture seriously, we need to confront the intrinsically relational nature of information. So, we need to ask some questions. If we are talking about one bit appearing in the world, what system's uncertainty is reduced by this bit of information, and uncertainty regarding what? Is it appropriate to say that not only can a particle's spin or momentum be uncertain, but also its identity altogether? Is it possible for *all* properties specifying a particle to be uncertain to *all* observers, and therefore, for that particle really to not exist in the world in any manner of speaking? These questions need to be addressed before "it from bit" can be developed into a coherent theory.

Wheeler's and Rovelli's ideas can come into alignment if we extend relational quantum mechanics, using Rovelli's own fundamental principles, so that it describes the relations between *informational systems only* — as opposed to systems of stuff from which information can be derived. The title of my essay that this book is based on, "Toward an Informational Mechanics," reflects the desire for such a theory. A theory of informational mechanics would mathematically describe how information operates in the world. If properly formalized, such a theory could predict everything that conventional quantum mechanics predicts, while also being a truly fundamental account of the bottom layer of reality. Understand that we are not trying to create a new quantum theory, or even add anything to the existing one. We are merely seeking to *generalize* Rovelli's theory in terms of informational relations between informational systems, rather than informational relations between systems made of stuff. If we are successful, not only will we be able to predict and analyze the behavior of the matter and energy that derives from information, we'll also be able to apply the theory to other investigations that have proved difficult, such as the physics of the early universe, the origin of life, and even consciousness. That is the goal of the (future) theory of informational mechanics. I will develop one of its core principles in the remainder of this chapter. To get the ball rolling, I will use an analogy.

Information Building Upon Information

A ship in the middle of the ocean, sometime in the early 20th century, hits an iceberg and starts to sink. Frantically, someone in the ship's telegraph office sends out a distress signal:

$$\bullet \bullet \bullet \; — \; — \; — \; \bullet \bullet \bullet \qquad \bullet \bullet \bullet \; — \; — \; — \; \bullet \bullet \bullet \qquad \bullet \bullet \bullet \; — \; — \; — \; \bullet \bullet \bullet$$

These are just dots and dashes encoded into radio waves. In fact, to anyone who hadn't learned Morse code, this signal would be meaningless noise. It wouldn't be much different for a telegraph operator who knew Morse code, without knowing about the SOS distress signal — he'd hear just a series of letters, and they wouldn't mean anything. However, if the receiver holds information that associates the dots and the dashes with emergency, suddenly the series of letters becomes meaningful ("a vessel is in distress"). If he furthermore knows the maximum range at which he is likely to receive telegraph signals, the content of the message rises higher ("a vessel within 100 miles is in distress"). Finally, if he knows that a passenger ship is within this distance, and that it has 2,000 souls aboard, the message becomes extremely urgent. With each additional piece of old information added into the mix, the informational value of the new information (the dots and dashes) goes up. In other words, an accumulation of information from the past — knowledge of Morse code, knowledge of distress signals, etc. — *constrains* how the new information gets interpreted, and the result is a correspondingly greater reduction in the telegraph receiver's uncertainty about what is going on.

For the telegraph operator, what might be interpreted as someone transmitting a repeated pattern for no reason becomes sharpened into a dramatic and important message. The dots and dashes, interpreted within the context of old or legacy information, become very meaningful, with a high informational content (i.e., much uncertainty reduction) — at least with respect to the person who holds all of that information. For some other receiver with less legacy information to apply, the same message would have a much lower informational content, as uncertainty remains high. Notice the similarities with relational quantum mechanics, where different observers can give different descriptions of the same thing, without those descriptions being inconsistent with each other. One sender-receiver relation may constitute a lot of information ("a ship with 2,000 passengers is in distress" — the receiver's uncertainty about the situation is greatly reduced), while another relation may constitute very little information ("someone is transmitting a repeated pattern, and I have no idea what that's about").

The signal represents an informational relation, between the sender and the receiver. But the receiver also brings certain other informational relations to the table or desk, which he has learned as a professional telegraph operator. For example, he probably knows certain equivalencies or identity relations, such as (• • • = S), (— — — = O), and (SOS SOS... = distress signal). These pieces of information can reduce his uncertainty about what letter is represented by three dots, for instance. Thus these legacy relations are clearly informational, relative to him, or to anyone who knows them.

Something interesting happens with all these relations in the receiver's mind, as the signal is coming across. Hearing three dots, he recalls from memory the identity relation (• • • = S), and writes down the letter "S." Hearing three dashes, he similarly recalls the appropriate identity relation and writes "O." This continues for a short while, and he creates a new system of informational relations that looks something like this: SOS SOS SOS. Of course, that system could also be expressed with all of the individual identity relations written out: (• • • = S)(— — — = O)(• • • = S), and so on. Notice that this larger system is actually a *relation of relations*. Finally, the telegraph operator recalls another identity relation (SOS SOS... = distress), and creates in his mind an even larger system still, the relation between (SOS SOS SOS) and (SOS SOS... = distress). Noticing that this larger system is yet another identity relation, alarm bells go off. The information contained in those measly dots and dashes becomes elevated to a matter of life or death involving thousands of people. (See **Fig. 4**, next page.)

With this example, we see how systems of informational relations can be nested within other systems of informational relations, to create a larger system that's a complex of the smaller systems. And the larger system can be nested within a larger system still. In this case, legacy information — which we sometimes call knowledge — is assembled into systems that are useful for interpreting the incoming Morse code signal. The telegraph operator's job is to interpret the new information. That means, using his knowledge and other available information to turn telegraph signals from messages that seem useless (dots and dashes over the radio) into ones that may be extremely detailed and specific. Legacy information, such as the equivalencies of Morse code and details about other ships in the region, serves as a kind of lens that focuses and sharpens the new, incoming information. Like a telescope that you can insert more and more lenses into, the picture becomes increasingly detailed and rich with each piece of legacy information that gets applied.

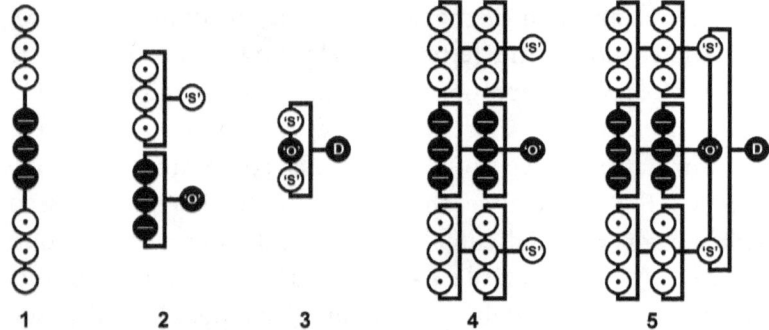

Fig. 4. Raw information such as a string of dots and dashes (1) is meaningless. However, a person may have certain identity relations, such as knowledge of letter equivalents (2) and the distress signal (3). Combining (1) and (2) allows the string to be decoded into letters (4), and by combining (4) with (3), the meaning of the string (5). For simplicity, I have omitted the message's receiver from the diagrams. Think of yourself as the receiver, as you look at the diagrams and enter into informational relations with each of them.

The story with the telegraph operator may not end there. If after the SOS signal, he receives Morse code for the word FIRE (not necessarily a distress signal in itself), then those four letters get a super-boost of informational value — not only from the Morse-letter equivalencies known by the operator, but *also* from the SOS distress signal having been previously received. The SOS has persisted in the world (in the operator's memory, or on his sheet of paper) to become its own piece of legacy information, a lens that focuses the informational content of the word FIRE. Now, the meaning of the message is the sharpest it's ever been: "A nearby ship with 2,000 passengers is on fire."

This provides a glimpse into how a universe that's composed of information could evolve from very simple, containing very little information, to very complex, containing a great deal of information. If legacy information sharpens or increases the uncertainty-reducing power of new information, then when a system accumulates more and more information, the power of any new arriving bit to reduce that system's uncertainty will grow higher and higher. For a system that has accumulated information for, say, *billions of years*, the detail provided by the arrival of any new bits — for example, in the form of photons registered by a space telescope named Hubble — could be truly spectacular.

The Role of Context

At this point, you may feel that it's a stretch to spin a tale about a ship sinking, and claim that the story can shed insight upon things like electrons and atoms, let alone the ultimate nature of reality. But even though the story is an analogy, it does illustrate a governing, universal principle of the world: Information gains value when it's combined with other information. This is the role of informational *context*. Context is not a word that shows up often in quantum literature, although some have suggested that context does have a fundamental importance to the nature of things.[3] That is what informational mechanics aims to do: to provide a formal, mathematical description of the effect provided by legacy information (i.e., context) upon new information arriving in the world, relative to the systems involved.

If you're still not convinced that old information sharpens new information, let's return to the experiments. Consider the double-slit experiment with individual photons, where our task is to determine the wavelength of light by letting photons slowly

---

3 As the physicist Vlatko Vedral (author of *Decoding Reality: The Universe As Quantum Information*) has said, "We need some kind of 'relative information' concept, information that is not only dependent upon the probability, but also its context."

accumulate on the screen (page 100). We begin with information about the experimental setup: We know the distance between the slits, and we know where we placed the screen. This physical setup contains informational relations. (Arguably, it *consists* of these relations!) For example, there is the ratio of those two separation distances: the distance between the slits, vs. from the slits to the screen. This is a unitless number that can be calculated by measuring the distances in terms of any arbitrary unit of length, and then dividing the distances, which gets rid of the units. But this ratio information alone is insufficient to determine the wavelength of the light — we haven't even registered the first photon yet. To determine the wavelength, we need to measure the distance between the interference fringes, *relative to the ratio of distances in the experimental setup.* (See **Fig. 5** at right.) The first photon that arrives on the screen, on its own, provides no information on the fringe distance. For that, we need a lot of photons. Given an accumulation of photons, in other words all of the photons considered *in context of one another*, we have the last piece of information necessary to determine wavelength.

According to the relevant equation, the wavelength equals the distance between the fringes multiplied by the experimental setup ratio (the slit separation divided by the distance to the screen). If we know that the slits are separated by 0.5mm, and it's 5000mm to the screen, then the setup ratio is 1/10,000th, or .0001. This is legacy information that we will combine with the *new* information on the distance between the fringes, to find the wavelength of the light. Suppose we measure the fringes to be 5mm apart. Starting with the experimental-setup ratio of .0001, and multiplying it by the fringe measurement that we reached by considering many photons in the context of one another, we find that the wavelength is .0005mm, or 500 nanometers — a bluish-green. Notice that measuring the fringe distance in millimeters results in the wavelength being expressed in millimeters. They are the units of scale in the coordinate system that we needed to specify, in order to make measurements that have any meaning.

Of course, if any of our physical distance measurements is uncertain, that will introduce uncertainty into the relational system, and its informational power will go down correspondingly. If we can measure the distance between the fringes only to an accuracy of plus-or-minus 20%, that means we'll have a corresponding uncertainty in the wavelength we are seeking. In this calculation, we can see relations nested in other relations: Each known distance in the setup is, itself, a relation with a relation, involving for example the locations of the slits vs. the screen, as well as the two ends of a standard measuring-stick (as described on pages 101–105 and in **Fig. 5** below). These distances are considered as a ratio. That setup ratio provides the context in which we interpret the distance relation between the fringes on the screen, which, in turn, is arrived at by considering all of the photon spots in context of each other. This quantum experiment is not that different from the SOS analogy: In both cases, legacy informational relations assemble into systems, and new information is sharpened by the assembly of these systems into larger systems.

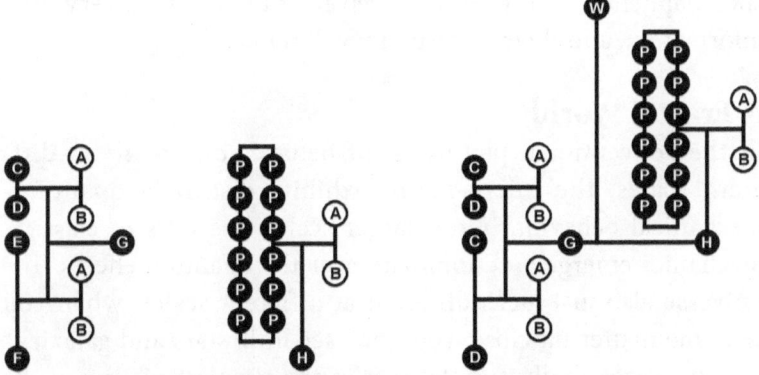

Fig. 5. At left, the double-slit setup comprises the measurement unit **AB**, slit separation **CD**, and distance to the screen **EF**; the ratio between **CD** and **EF** is the setup ratio **G**. At middle, many photons (**P**) are considered together relative to the measurement unit **AB**, to measure the fringe distance **H**. At right, the relations **G** and **H** are combined to determine the wavelength **W**. (Again, the experimenter is omitted from the diagrams for simplicity.)

To take this argument another step further, consider the experiment at the beginning of Chapter 1: We heard a click. If we have no contextual information in which to interpret that click, our uncertainty regarding what happened isn't reduced at all, beyond "*something* happened." But, with every piece of legacy information added into the mix, the information conveyed in the relation between click and experimenter becomes sharper. If we have information about boundaries in space and time (i.e., where and when the experiment took place, relative to known locations/times), as well as information regarding the properties of the sample (uranium?), as well as knowledge about Geiger counters and what they do, *only then* may we arrive at a statement as informationally rich as, "Between 12:00 and 1:00 PM today, in my lab, an alpha particle was detected." Notice that if you want to convey meaningful information about the experiment to someone else, your description must embed contextual information about space and time, etc. It's the only way your news can reduce the other person's uncertainty about much of anything. You need to spell everything out; merely saying, "A click happened" will give the receiver of the message very little information, and their uncertainty will remain high.

A Fractal World

In the conventional picture, stuff behaves differently at different scales; the micro-realm exhibits distinctly quantum-mechanical behavior, but at larger scales, the laws of classical mechanics emerge and dominate over the quantum effects. The universe also just *looks* different at different scales; when you examine matter up close, you don't see little stars and galaxies.[4] Occam's razor might suggest that in the simplest picture of the universe, everything should ultimately operate exactly the same at all scales and in every realm of description. That isn't what we

---

4 Interestingly, though, empty space dominates both atoms and the largest-scale structures in the universe.

see in the stuff picture. By contrast, in other ways nature often displays *fractal* patterns, where structures are nested within similar structures that repeat at multiple scales, the classic examples being the leaves of a fern or the branches of a tree. One of the more elegant implications of the simplest-case scenario is that the universe is ultimately fractal.

Fractals are especially prevalent in living things. It's simpler and more efficient for DNA to encode a few basic rules for building structures out of other structures, rather than to encode structural blueprints for every individual cell of every anatomical feature (which DNA doesn't do). If the ultimate nature of things is informational, then this same principle might be expected to be universal across all scales. When we speak of information in the human realm or the bacteria realm, the governing principles may be the same as for information in the atomic realm, the technological realm, and the galactic realm.

Most dramatically, according to this informational picture, the biological realm is fractal through and through. Each organism consists of complex systems nested in other complex systems, down to the cellular and sub-cellular level. In particular, DNA is an excellent example of information nested in information: Structures called nucleobases — which are like bits, but base-four instead of binary[5] — form base pairs, which are nested inside genes, which are nested inside chromosomes, which are nested inside the genome. And, it keeps on going: Some social organisms, such as termites and honeybees, can be considered components of larger superorganisms, whose behavior is driven by complex, emergent information. As Chapter 8 will explain, the ultimate superorganism is a *super-observer* of the biggest informational system of them all — and we call that system the universe.

---

5 In DNA, a nucleobase can be A, C, G, or T, in the way that a bit can be 0 or 1: One information scheme is base-four, and the other is base-two. This is my whimsical interpretation of the number 42 as the "Answer to the Ultimate Question of Life, the Universe, and Everything" from Douglas Adams' *The Hitchhiker's Guide to the Galaxy*.

The Informational–Relational Universe

One enlightening concept shows up from time to time in foundational physics and the philosophy of physics: *There cannot be a complete, exact description of the universe as a whole.* Lee Smolin points out that if a theoretical observer could measure the entire universe, gaining this information would physically reconfigure the observer — and, being a part of the universe, the observer would then have to re-measure the entire universe. Such an endless process could never lead to a complete description. This is how Carlo Rovelli puts it: "There is no description of the universe *in-toto*, only a quantum-interrelated net of partial descriptions." Recall the analogy of the network of friends with debts (page 105): No individual friend is involved in all of the debts. There are six relations, but each friend is privy to only three of them; thus each friend has only a partial description of the entire system. And, as we saw, even though each friend owes or is owed money, if you as an external onlooker add all of the relations, you get zero. As I suggested in that chapter, this is how some scientists view the universe in the ultimate scheme of things. It's as if the universe as a whole doesn't exist — that only its parts exist! This is consistent with Rovelli's statement that "There is neither an absolute state of the system, nor absolute properties that the system has at a certain time." To describe the universe as a whole is to describe its absolute properties. And there are no such properties.

In the next chapter, we'll bring in the many-worlds interpretation, and see how our universe may be one possibility out of many possible universes. After that, we'll be able to see the big picture of how our universe is built out of a complex, self-generating set of informational relations. Our deepest questions about existence, from quantum mechanics to consciousness to the Fermi paradox, may be variations on the same Big Question, or different aspects of it. It's a question that we've never really asked ourselves in full, let alone found the answer to. Perhaps now we finally have.

Things to Remember From Chapter 5

• The debate over how to interpret quantum mechanics has been going on since the theory was first formulated.

• Carlo Rovelli's relational quantum mechanics interprets variable properties of things not as absolute values, but as relations with some observer, whatever that might be. Different observers can thereby obtain different measurements of the same thing, without those measurements being inconsistent with each other.

• Relational quantum mechanics can be unified with John Wheeler's it-from-bit conjecture, by proposing that even the properties identifying particles are not absolute values, but instead are informational relations with some observer.

• The result is a fundamentally informational, fully relational interpretation of quantum mechanics, *informational mechanics*.

• According to informational mechanics, contextual (or old) information sharpens new information, increasing its uncertainty-reducing power. Reference frames (discussed in Chapter 3) are an example of contextual information. Knowledge is another.

• The principles of informational mechanics apply equally and consistently to all realms and scales, up to the scale of the entire universe. However, since no observer can have a relation with the entire universe, nothing can be said about the universe as a whole. Only its parts or subsystems can be described.

CHAPTER 6

MULTIPLE UNIVERSES, POSSIBLE UNIVERSES

H ave you ever thought about the unlikely chain of events that has led to you being here right now to read this book?

Maybe not. If you had, your brain might have exploded, and you'd probably remember that.

I'm not talking about events in your life or decisions you've made, although those do factor in. I'm not talking about your parents happening to meet by chance, or a sperm cell reaching an egg, although those factor in as well. I'm talking about a chain of events stretching back hundreds, thousands, *millions*, even *hundreds of millions of years* and beyond — a chain that has ended, thus far, with you being here today to think about it.

What events exactly? Well, consider this profound observation by the biologist Richard Dawkins: "Not a single one of your ancestors died young. They all copulated at least once." You find yourself today at the end of a continuous, unbroken lineage of parents producing offspring, and those offspring producing offspring, and so on and so on, that goes *all the way back* to the earliest reproducing life on Earth. There is no ambiguity about this idea: Starting with the earliest common ancestor of all living things, several *billion* years ago, each individual organism that derived from this ultimate ancestor, and which was part of the lineage eventually leading to you, had to survive long enough to reproduce. It also had to find a partner with which it could copulate,[1] and that copulation had to be successful at producing living offspring. Then at least one of the offspring had to elude predators, diseases, and the elements long enough to reproduce in turn. It's remarkable enough that this can happen even once, that new life can be created out of old life. But in order for you to be here today, that process of survival and reproduction, often against steep odds, had to happen tens of millions, hundreds of millions, or billions of times. And not just that many times total — that many times *in a row!* All it would have taken was one fish snapping up an egg in a flash, or one volcanic rock landing in the wrong place, or one bird of prey swooping in, or one epidemic, etc., *at any moment* during the past billion-plus years, and the chain would have been broken and you wouldn't be here. Maybe none of us would.

This astonishing unbroken chain is like a delicate, wispy thread of spider-silk stretching across the light-years through countless star systems, and managing to persist intact despite asteroids and rogue comets and high-energy particles and everything else the Cosmos can throw at it. We're talking about something like Joe DiMaggio's famous hitting streak being not just 56 games long, but 56 *million* games long, at least. Without DiMaggio making a single out!

Believe it or not, the outrageous unlikelihood of your existence doesn't stop there. No one dictated that there had to be life on Earth capable of reproducing in the first place. No one dictated that there had to be an Earth, friendly to life, in the "just right" habitable zone of the Solar System, with liquid water and conditions amenable to complex chemistry. No one dictated that there had to be gravity to collect matter into stars and planets. No one dictated that there had to be nuclear forces and other parameters that allow the existence of stable atoms and molecules of matter. No one dictated that there had to be anything at all.

---

1 Biologists believe that sexual reproduction evolved about one billion years ago. Prior to that, organisms didn't have to find a mate — but they still had to survive.

Now, people who approach the world with a theistic mind-set aren't impressed by this line of thinking. For them, your ancestors' survival, and the creation of life, and the conducive environment of the universe for matter and life, are all a part of God's Plan. This teleological or purpose-oriented viewpoint says that our existence is somehow thought out and designed with intent, and to believe otherwise is to say that we are here because of some kind of "accident." Creationists in particular love to use that word, which evokes a kind of disastrous outcome. The world is beautiful, so it can't *possibly* be an accident! — or so they would like you to think. But secular thinkers prefer to give our existence a more positive spin, saying that we won the cosmic lottery, or something to that effect.

I will not battle theists and creationists here; that is a topic for another book. But if you don't outright accept the conjecture that the universe and Earthly life have been guided by an intelligent, all-knowing, omnipotent hand, the questions remain: Why is the universe the way it is? Why are its laws such that they allow complex chemistry and life (or matter for that matter)? The universe might have contained nothing at all from the beginning — so why does it contain anything?

This concerns a controversy in science called *fine-tuning*. Depending upon whom you ask, there are several (or many) numbers related to the laws of physics which, if they were slightly different, would lead to the universe being nothing like the one we see, with stable matter, gravity, etc. Some describe the universe as being on a "knife's edge" where these parameters balance just so; others assert that there is no such fine-tuning. Personally, I believe that fine-tuning is simply a fact of nature, and that many who criticize it do so not from a truly objective analysis, but rather, largely out of discomfort with its theistic or teleological implications. After all, if you say that the universe is fine-tuned, that implies the existence of a tuner. While in Newton's time it was perfectly okay to assume an invisible, untestable hand of God, that is simply a non-starter in modern science. But fine-tuning seems to be true in the same way that the unbroken chain of life leading to you is true — clearly, it actually happened, because you *are* here. It then becomes our job to weigh the various viewpoints and figure out the best explanation for why the universe turned out the way that it has.

Viewed in retrospect, it may seem extremely unlikely that a chain of parents-and-offspring could last, uninterrupted, for billions of years to lead to you. But that isn't really a fair analysis. If there had been a break in the chain — your grandmother died of polio, or some proto-lemur got eaten 70 million years ago, or what have you — you wouldn't be here to ask the question. This relates to a concept in the philosophy of science known as the *anthropic principle.* It states that the universe appears to be fine-tuned to support the existence of matter and life (and eventually consciousness, and you) simply because there *is* matter, life, and consciousness, and you, to ponder this state of affairs. If there weren't, no one would be around to notice that the universe *doesn't* support matter and life.

On its own, though, and without appealing to an intelligent designer or something similar, the anthropic principle is unsatisfying. It suggests that if there is only one universe, with one set of physical laws — which is what ordinary observations seem to suggest — we must be astonishingly fortunate that these laws happen to support stable matter, complex chemistry, and so on. It's as if the unbroken chain of life leading to you was the only such lineage that ever existed on Earth, and through mind-boggling happenstance, this one thread managed to keep going for billions of years. Obviously that's not the case; most of the reproductive threads that have appeared throughout the history of life did, in fact, end with a death before copulation somewhere along the line. Only a handful have made it to today: specifically, those ending with the organisms that are alive right now. So, even if your mind is blown by realizing your unlikely existence at the end of that long chain, the anthropic principle provides a satisfactory explanation: There were trillions of threads that did not make it. But you can't say the same thing in the case of a solitary universe that might be in any configuration, and happens to be in this one. The anthropic principle, by itself, doesn't help us understand why we happen to live in a universe with just the right kind of order to support stable matter, life, and consciousness.

According to the simplest-case scenario — and supported by a Stephen Hawking theory, as we'll see in the next chapter — fine-tuning is largely a matter of perspective. It's an inevitable consequence of a universe made of information, developing in complexity to the point where it includes complex intelligent observers able to make such realizations. And, that doesn't happen out of any kind of accident. Consider a craps player rolling dice: Is it an accident that a certain number of dice rolls ended up the way they did? Well, that would be a strange way of describing it. Each roll could have gone various ways, and the series of rolls as a whole could have gone in many other ways — but it didn't, and that's all that needs to be said. Is it an accident that all of your ancestors survived long enough to reproduce? Again, that would be a strange way to put it. Each one of them could have lived or died, in the same manner that the dice could go this way or that way. It just so happens that one particular thread of ancestry leads to you being alive today. That makes your existence not an accident, but it's also not due to the intentional guidance of some overseeing deity. The same can be said of the universe.

The Multiverse Saves Fine Tuning

To accompany the anthropic principle, we need a cosmological equivalent of the trillions of reproductive threads that have existed on Earth, almost of all of which didn't make it to today. If we knew that there were many, many universes out there, each a little different from the next — and perhaps some of them not supporting complex chemistry or even matter at all — then we shouldn't be too surprised that one is just right for matter and life, and here we are living in it. It's like winning the Powerball Lottery jackpot: Yes, you were lucky to win, but think about all of the people who didn't win. They aren't sitting around reflecting on their fortune; it's just another ordinary day for them. Anyone could have won, so why not you? That's the kind of thing the anthropic principle needs, in order to explain how the universe can seem to be fine-tuned for matter and life.

Enter the *multiverse*. According to multiverse theories of physics and cosmology, there exist a nearly infinite or perhaps infinite number of universes, only one of which we observe as our universe. These universes can have very different characteristics from one to the next. If there are many universes, the anthropic principle can be properly invoked: The unlikely fine-tuning of one universe (ours) can be explained by way of all the universes that *aren't* fine-tuned. It makes no difference if a universe like ours is rare. It's like the trillions of threads of life that have lived and died on Earth, where it takes only one to make its way across the eons to lead to your conscious self, to explain how you could be sitting here contemplating it all. No matter how unlikely such a thing might be.

Multiverse theories are controversial. Some physicists, including Stephen Hawking, Max Tegmark, and Neil deGrasse Tyson, believe that different universes must exist in some sense, to explain how ours ended up the way it did. Others, including Roger Penrose, George Ellis, and Paul Davies, have difficulties with the proposal of a multiverse. Davies has been particularly critical, arguing in a *New York Times* op-ed that any multiverse theory amounts to a theological-type leap of *faith* in modern science. Since the multiverse can be neither observed, nor proved, nor disproved (at least at present), Davies argues that the explanation is no better than asserting that the universe was intentionally fine-tuned by an external agent such as an intelligent designer. Faith in mainstream science? Well, you can imagine the reaction. Some scientists almost seemed to take

Davies' challenge as a personal attack. Personally I think he has a valid point, although some multiverse theories seem to be more faithful than others.

In an influential paper, Max Tegmark introduced four types of multiverses. They're worth reviewing to understand the similarities and differences among various multiverse theories.

A **Level I multiverse** emerges from the concept of cosmic inflation, where at the Big Bang, the one-and-only universe expanded at an astonishing rate — so much so that there are countless local regions or "Hubble bubbles" so far away, we could not observe them, nor could we ever observe them. Although these regions all share the same physical laws and constants, the configurations of matter and energy are different from region to region. So, if someone could travel much, much faster than light, eventually they would find another region exactly like our observable universe, even including human beings exactly like us. This idea isn't new, and I remember being blown away the first time I came across it as a kid.[2] Although the regions can be thought of as individual universes, in the sense that each region can only be observed by inhabitants inside, it doesn't help with the fine-tuning question, since all of the regions would presumably be "tuned" the same and would contain the same kinds of matter, just clumped up differently.

A **Level II multiverse** takes the concept a step further: There are other "bubbles" out there that may be governed by laws similar to ours, but the physical constants are different,[3] and these other universes may have different kinds of matter (or no matter) and perhaps even different numbers of dimensions.

---

2 Only slightly related but funny: On an early *Saturday Night Live*, Father Guido Sarducci (played by Don Novello) reported that there was another planet on the other side of the Sun that was exactly like Earth in every single way, except for one detail: "They eat-a corn on the cob like-a this" (holding the corn cob vertically).

3 As they would be measured by us, if such a thing were possible — or perhaps as they would be measured by intelligent inhabitants of that world using mathematics similar to ours, if that were possible.

Level II thereby facilitates the anthropic explanation for why our universe appears to be fine-tuned: There are many universes whose physical parameters do not support matter and life — but a few do, and we live in one of them.

The **Level III multiverse** gets even more interesting. Physically, it has the same features as Level II, but this kind of multiverse is governed by quantum mechanics in the form of the many-worlds interpretation (see pages 118–119). This means that even within a single region of space, as it were, there exist different versions of universes with different configurations of matter, different outcomes of events, etc. This is the classic idea of parallel universes from science fiction. Tegmark points out that a similar situation of parallel-type universes might be found even within a Level I multiverse — it's just that you'd have to go a very long distance to reach one. A universe containing an identical copy of you, for example, but with a different outcome for an atomic experiment, would be extremely far away.

Finally, there is the **Level IV multiverse**. This is just like Level III with the parallel universes splitting off from each other, but individual universes are ultimately mathematical structures, and the Level IV multiverse can contain structures (i.e., universes) that are wildly different from each other. There may be universes with entirely different laws of physics and even different mathematics. Tegmark points out that this is the ultimate extension of the multiverse concept, encompassing any other type of multiverse that could be proposed; there cannot be a Level V multiverse, for example.

Each level of Tegmark's hierarchy gets us closer to a satisfying resolution of the fine-tuning question. In Level IV, *every* kind of universe with *every* variation of physical laws, constants, and mathematics exists in some sense, so there should be no surprise that at least one contains matter and life — no matter how balanced-on-the-knife's-edge that world may appear to be. This would answer a deep question that John Wheeler had:

Why do *these particular equations* govern the universe, and not others? If there exists a Level IV multiverse, then there are countless other universes governed by their own mathematics, so that pretty much answers Wheeler's question.

But even if we accept Tegmark's hypothesis that there exist other universes consisting of entirely different mathematical structures, a burning question remains: *How does each universe get to be that way?* Unless we wish to invoke the intelligent-designer hypothesis for the various universes in the multiverse (and you won't find me doing that), we need to address how these structures might come about.

The Plinko Analogy

On the television game show *The Price Is Right*, there is a game called "Plinko." A chip is dropped into a vertically oriented box with horizontal pegs in it, and the chip bounces around the pegs as it makes its way to the bottom. It's an excellent way to visualize probabilistic behavior; in fact, a similar device was invented in the 1880s by Sir Francis Galton to investigate probability.[4] If you drop a chip into a Plinko box, the chip may take any of a number of paths to the bottom. Whenever the chip falls between two pegs, it hits a peg underneath and must "decide" whether to go left or right. In this manner, each row of pegs that the chip encounters is like a coin toss, where we would expect the chip to jog to the left or to the right with equal probability. Stated another way, it's as if the chip is presented with a series of yes-or-no questions, and the chip chooses an answer every time, effectively at random.

---

4 Technically, the falling chip is exhibiting chaotic behavior: Rather than its path being determined strictly by the laws of probability, the path depends upon the precise way it was dropped, as well as minute factors such as air currents, dust, the exact shape of the chip, unevenness of the pegs, etc. Still, the path is sufficiently unpredictable and unreproducible, and the outcome closely enough simulates probability laws, that we can think of the chip's path as being determined by probability. (The same is true of a well-performed coin-flip or dice-roll.)

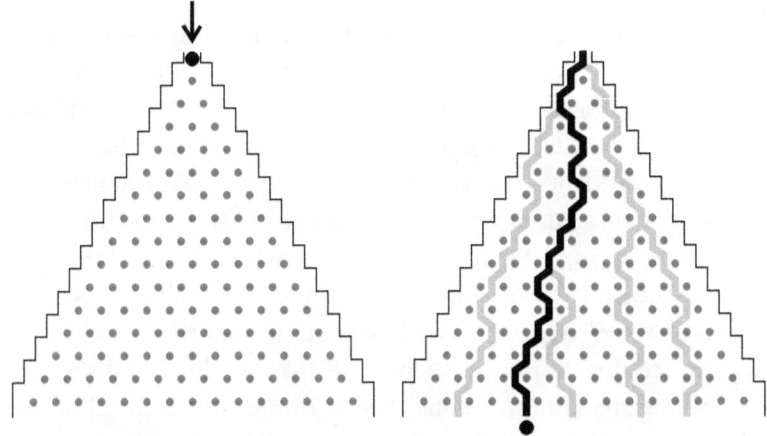

Fig. 1. A chip dropped into the top of a Plinko box (left) can take many possible paths (right) — 32,768 in this case, five of which are shown. In the end, though, we see it take only one path (black). The others can be thought of as alternate histories that didn't happen in this universe.

Imagine a large Plinko box. You could sit there and tediously trace out every possible pathway that a chip could take, and count them all, but math makes things like this easier. The number of possible paths from the top to the bottom is equal to 2^n (two to the power of n), where n is the number of rows. Just one row of pegs (i.e., the chip encounters only one peg) and there are two possible paths, left and right. Add another row and there are four possible paths. The number of paths goes up quickly as rows are added — with just ten rows, a chip can take 1,024 possible paths, and with 15 rows, there are 32,768. But, before you drop the chip, all of these are only *potential* paths, and the probability of the chip taking any one path is equal to the probability of any other path. In other words, with 15 rows of pegs, any one of the potential paths has a 1/32,768 chance of being realized.[5]

---

5 That's not to say the chip has an equal chance of winding up in any of the positions at the bottom. Statistically, more paths terminate near the middle than near the edges, so a chip is more likely to finish toward the center than at an extreme edge. This is how Galton used his device to illustrate the statistical phenomenon known as regression toward the mean.

When you think about the path that your life has taken, or the path of humanity, or life in general or even the Earth, that path is like one of the pathways through a very large Plinko box. Things could have gone differently; you might have won the Powerball Lottery, Archduke Franz Ferdinand's assassin might have missed, or the Earth might have been sterilized by a gamma-ray burst from a nearby star, which is an extremely unlikely though potential event. But these things didn't happen, and our history jogged this way rather than that way.

Notice the tie-ins between the Plinko analogy and themes we have discussed in this book: Before dropping a chip, all of the paths are possibilities. We are uncertain of which path the chip will take — we have no information about it — and the one that will get realized is in our future. If we covered the face of the box so we couldn't see inside, and then dropped a chip, we would still be uncertain of the path the chip had taken. Without receiving any information on what it had done, all paths remain possible, and the chip's actual path is *still* in our informational future, from our perspective (i.e., relative to us). If we then opened the box to see the position of the chip at the bottom, we would get *some* information on the path the chip had taken, but not much. If the chip wound up near the far left edge, this information might reduce our uncertainty from 32,768 possible paths to only a few, whereas if the chip is in the middle, it might have taken hundreds of different paths to get there. It's only if we watched the chip descend every step of the way, and we kept a record of the path, like left–left–right–left–right, etc., that we would have *complete* information on the path of the chip, and our uncertainty of the exact pathway would be reduced to zero. How much information would this constitute? One bit for each row. For a 15-row box, it would take 15 bits to specify which of 32,768 paths the chip took. You can imagine the results of experiments involving probability, like those discussed in Chapter 2, arising in a Plinko-like fashion as new information appears in the world that wasn't there before.

There's another tie-in with the Pinko analogy: When the chip has taken a particular path, this history then constrains what can subsequently happen to the chip. In no case could a chip falling along one side suddenly jump to the other side, for example. This would be like one particle's spin being measured in opposite directions along the same axis, or two humans producing mantelope offspring — there are laws of logic that constrain such things, and those laws dictate that certain courses of events just cannot happen.

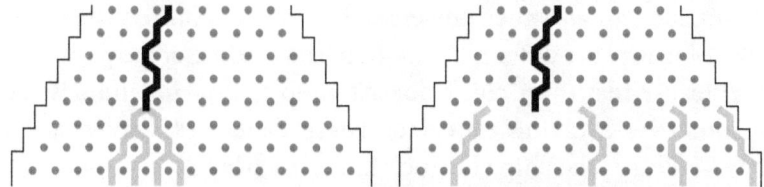

Fig. 2. A Plinko chip's history constrains what can happen subsequently: At left, there are only eight possible ways the chip can traverse the last three rows, four of which are shown. The paths at right are impossible.

Now at this point, I'd like to extend the analogy to the extreme, and imagine that the potential pathways inside the Plinko box are also like the various pathways a universe can take along its history. I'm talking about not just our own universe, but *any* universe, in the sense of Tegmark's Level IV multiverse. Perhaps all possible universes start out pretty much the same: This is analogous to the chip being dropped into the top of the box, where right from the start, things can go either this way, or that way. In this manner, John Wheeler's it-from-bit universe starts out with *only a single bit* — the simplest possible scenario, and all universes start out with this utter simplicity.[6]

6 Based on this paragraph, you might think that the simplest-case scenario is incompatible with Big Bang observations. It isn't, something that will be explained in the next chapter. Remember that in the scenario, matter (such as what came out of the Big Bang) derives from information and not vice-versa. Thus, the universe's informational beginnings may be very different, and far simpler, than its assumed material beginnings.

What happens after that is literally anyone's guess. Just as there are many pathways that a chip can take through a Plinko box, there are many potential pathways that a universe can take. The direction that our own universe took, including all of the tiny details, such as you not winning the Powerball Lottery, is one and exactly one of these pathways. Other pathways represent things that might have happened, but didn't. Some potential universes include a much luckier version of yourself, who hits the multi-state Powerball. In other universes, a meteor does not strike the Earth 66 million years ago, and humans do not rise to dominance. In yet other universes, the fundamental constants or laws of nature are different. Finally, perhaps some of the pathways are dead-ends; that "chip" doesn't keep falling indefinitely, as in our ever-evolving universe. These dead-end universes go nowhere, as if the chip got stuck in place early on.

The Ultimate Plinko Box — the ensemble of all potential pathways that could be taken — encompasses all potential worlds, every configuration that a universe could possibly have. The splitting parallel universes of the many-worlds interpretation and science fiction are in there, as are universes that aren't fine-tuned for matter and life, as are completely weird worlds governed by incomprehensibly different laws of nature. It's a Tegmark Level IV multiverse all the way.

In the Plinko analogy, the pathways are like trails of information: Every left-or-right jog of the chip is analogous to the choice nature might make, whether it's between a spin-up vs. a spin-down, or a yes vs. a no, or anything. Thus the Plinko analogy, with its multitude of pathways, is a good way to imagine the multiverse from an informational perspective, where each unique and highly specific individual pathway through the box is like a unique and highly specific informational universe. A tiny percentage of all possible universes are "fine-tuned" for matter and life, but that's still a staggeringly large number. And, just *one* of them is the universe you find yourself in when your eye reaches the end of this sentence.

Hypothetical Ontologies

The simplest-case scenario suggests that all potential universes are fundamentally informational. But information is always relative to something (Chapter 3). So, if the scenario is going to be consistent with itself, we have to ask: These bits that make up a universe, analogous to the lefts and rights in the Plinko box — *what* are they informational relative *to?* This gets at one of the deepest questions existence has to offer.

Information is the reduction of uncertainty for some party or parties. These benificiaries of uncertainty-reduction are typically called observers. But we observers get information only from our own universe. So, the bits that make up a universe must be informational *only to observers that are part of that universe.* This idea recalls the many-worlds interpretation: A universe appears to divide whenever something occurs that might occur some other way. It's just another jog along the Plinko pathway — and thereafter, the histories are different along the various branches. Observers in each branch might be aware of what happened, may have memories of what happened, or may have recorded the event on their cell phones. No matter which branch you might be observing from within, that branch is "the" universe, and the alternate branches are like the parallel universes familiar from science fiction.

Philosophers and physicists, including Tegmark in his multiverse paper, describe two ways of looking at this many-branched situation: There's the *bird's-eye view*, and there's also the *frog's eye view*. The bird's-eye view is how some kind of external overseer, like a god, would describe things. It's a description that's removed from the action, looking down upon things without actually being a part of them. It's the way we intuitively want to describe the world; we are used to describing things as being distinct from ourselves, whether it's a speck of dust or a chair or a distant galaxy. In observing something, we instinctively feel as if we're just collecting information about it. So when we try to describe the universe as a whole, this is the way we naturally want to go about it: The universe is a *thing*, and it has such-and-such properties. But quantum mechanics threw a monkey wrench into the assumption that we can observe and describe things objectively, without our act of observation becoming a part of the equation (the so-called observer effect commonly abused by New Age mystics). Recall Lee Smolin's point that measuring the entire universe would require endless re-measurement (page 138). In contrast, there is the frog's-eye view, where we describe the world *from within the world*: We are a part of the action, and the description includes ourselves, doing the describing, while acknowledging the fact that we could never make a complete, accurate description of the entire world from inside of it, even in principle. Thus, whereas the bird's-eye view attempts to describe the whole universe (or multiverse) from the external perspective of some all-seeing onlooker, the frog's-eye view can never take in the whole picture; it is only a partial description.[7]

Since we are unavoidably a part of our world, the frog's-eye perspective seems to be necessary to describe the nature of our universe. However, if we wish to imagine universes other than our own, and see how the universes fit together, we need to employ a bird's-eye perspective. We aren't a part of them, and we don't have any information on them; the pathways that those universes might be taking are completely uncertain to us. The course of events in any parallel universe, once it split off from ours, would be entirely uncertain — it would be *hypothetical* at best. There would be countless possibilities, all unknown to us and providing no information whatsoever, just like the possibilities for where a Plinko chip might go from any position in the box. This is especially true for universes that diverged from ours at a point much earlier in their history. Embedded as we are in our

---

7 Recall from Chapter 5 the point that Carlo Rovelli makes in his Relational Quantum Mechanics paper: "There is no description of the universe *in-toto*, only a quantum-interrelated net of partial descriptions."

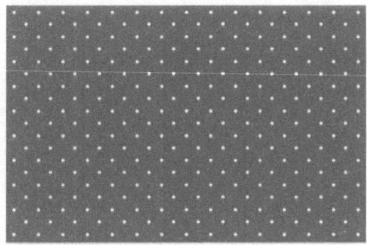 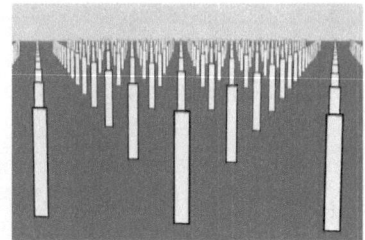

Fig. 3. One way to visualize the bird's-eye view is a field of regularly spaced posts, seen from above (left), which provides a generic look at the overall order. In the frog's-eye view of the same scene (right), the perspective comes into play, and specific patterns emerge in relation to the observer.

frog's-eye view of our own world, any information constituting another universe could reduce uncertainty *only for observers contained within that world* — that's the most we can say. Another universe's potential pathways are all hypothetical from our perspective.

But now, imagine trying to describe the course of events in our own universe, from the perspective of an intelligent observer in some other universe, who has no information on our pathway (in the same manner that we have no information on theirs). From their perspective, the course of events in *our* universe would be hypothetical. Only their own world would appear to them, while all other worlds, including ours, would be completely uncertain. This leads to a most interesting conclusion: If we want to try to describe *all* universes (including our own) in an objective, generic, bird's-eye-view manner — that is, if we wish to describe the components of the Level IV multiverse in a general and egalitarian fashion — the best we can do is say that they are all *possible* pathways. The true, fundamental state of being or ontology of any universe, described in the most generic "bird's-eye" sense, must be *just* that it is a potential, or a possibility. And that's all that's neccessary.

This explains how universes can exist in the first place: They necessarily *have* to exist, because there are always possibilities.

If you ask philosophers why there is something in the world rather than nothing, they might tell you that even if there were nothing, there would still exist the *possibility* of something, and such a possibility alone means that there isn't nothing.[8] The simplest-case scenario takes this metaphysical stance and expands it to say that the "something" we see in the world is just that: It exists because there is always the possibility of something.

In our particular case, that something has been taken to an extreme of informational richness and detail, as seen by observers within the world — us. At the fundamental level, though, our world is made not of particles of matter and energy born out of material from the Big Bang. Instead, it's a configuration of informational relations, one of many possible configurations, from which all of those particles emerge. And, all universes share this same ontological status: They are all *possibilities*, which you can think of as potential pathways in the multiverse's Ultimate Plinko Box.

I realize that this is extremely conceptual and abstract, so let me elaborate on what I mean by possible universes.

Networks of Numbers

Think of the change in your pocket. If you were to lay out the coins in a straight line, there are certain ways that they can be ordered. In each of these orderings, there exists a set of relations that the coins have with their adjacent neighbors. If they are ordered by value, a dime might have a nickel on one side and a quarter on the other. You don't have to actually line up the

---

8 Even in the *stuff* picture of the world, nothingness is problematic. The closest thing to nothing in physics is the empty space of the vacuum, which isn't really empty because it is roiling with quantum-field fluctuations, from which (according to the best cosmogeny theories) our universe sprang. In other words, even in the absence of the universe, there is still the vacuum, which contains the potential to produce anything and everything. Meanwhile, in mathematics the closest thing to nothing is the empty set, which is not nothing because it is, in fact, a set.

coins on a table in order for these relations to happen; you can imagine them being in that order, and you can imagine those particular relations, even while the coins are mixed up in your pocket. And of course, you don't have to imagine anything at all for those particular relations to be possible; the relations exist by the mere fact that the coins *could* be placed in that order. The particular relations (penny–nickel, nickel–dime, dime–quarter) are based on a particular ordering. Meanwhile, other sets of relations, such as (penny–dime, dime–quarter, quarter–nickel) also exist, for the exact same reason. Just by the fact of the coins existing, the various possible sets of their relations must exist, as well.

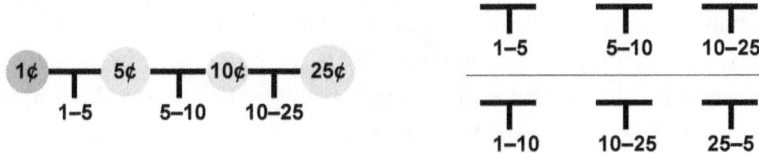

Fig. 4. Four coins laid out in order (left) exhibit relations between each pair of adjacent coins. However, these relations can exist even without the coins being physically laid out in that order, as can other possible relations among them (right).

Of course, that situation is based on a pocketful of material coins. To get closer to the idea of a universe made of information, let's make things more abstract. Instead of coins, think of numbers: 1, 2, 3, 4, and so on. Like coins, numbers can be ordered in various ways, either by writing them down, or thinking of them — but as with the coins, you don't have to do either of these things, or to get involved in any manner at all, in order for the numbers to have *possible* arrangements or sets of relations. There's 1–2–3–4, and 1–3–2–4, as well as 1–2–3–4–5 and so on. Other arrangement patterns are also possible, even if these patterns are more complex and the numbers appear more than once — for example, the pattern 1–3–2–4–3–5–4–6, etc. Furthermore, no one said that these arrangements have to be linear; numbers can be arranged in closed or ring-like patterns, including with diagonals, like the network of debt relations among the four friends in Chapter 4.

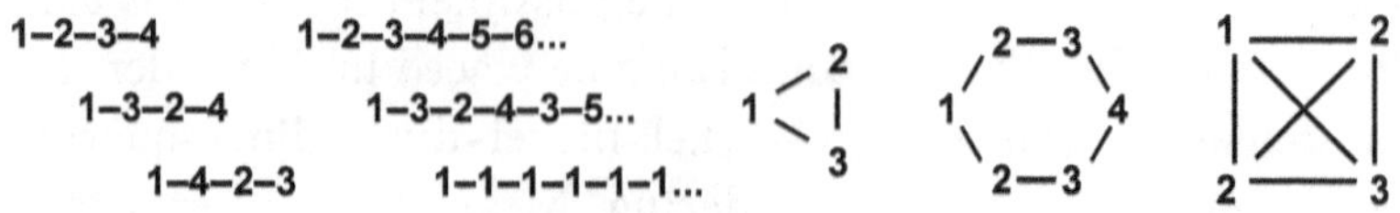

Fig. 5. A handful of the relations that can exist among natural numbers.

Some may argue that this argument is faulty, that numbers are invented by man and are not a part of the natural world. That's wrong. The natural numbers are called that for a reason. In Chapter 4, I mentioned that every atom of the element lithium has, by definition, three and exactly three protons in its nucleus. Four protons and it's not lithium anymore; it's beryllium. Also, any sample of matter could be observed to contain a specific, whole number of atoms, these atoms being countable in principle.[9] The natural numbers are abstract things when they aren't written down, but they are features of the world nonetheless.

We just have to go one more step in order to reach the bedrock of reality, and see, finally, how John Wheeler's it-from-bit universe might work. Instead of coins or a bunch of numbers, imagine just two numbers. Or two letters. Or two colors, or states of being, or two of anything you might want to use — as long as there are two and they are different, the words you choose to think about them are arbitrary. In my diagrams, I use black and white dots, but they might as well be A's and B's, or yes's and no's, or humans and antelopes. It makes no difference.

9 If you don't like that I am appealing to a *stuff* description of nature (protons, atoms, etc.), to defend the natural existence of numbers, perhaps you would prefer philosopher/ mathematician Gottlob Frege's approach: The number of things in the world not identical to themselves is 0. The number of such numbers is 1 — there is only one number that describes the number of things not identical to themselves, and that number is 0. This process can be reiterated to produce any natural number you desire.

Now, let's see what kind of possible "universes" we can build from the relations between two dots. They can occur in a series, as with the examples of coins and numbers (see **Fig. 6A**). They can be ring-like or repeating and periodic (**Fig. 6B**). Also, thinking back to Chapter 5's story about the sinking ship and the SOS signal, and what it demonstrates about information in general: Relations can be nested, such that relations involve other relations (**Fig. 6C**). Finally, there can be combinations of these structures, with progressive levels of complexity (**Fig. 6D**).

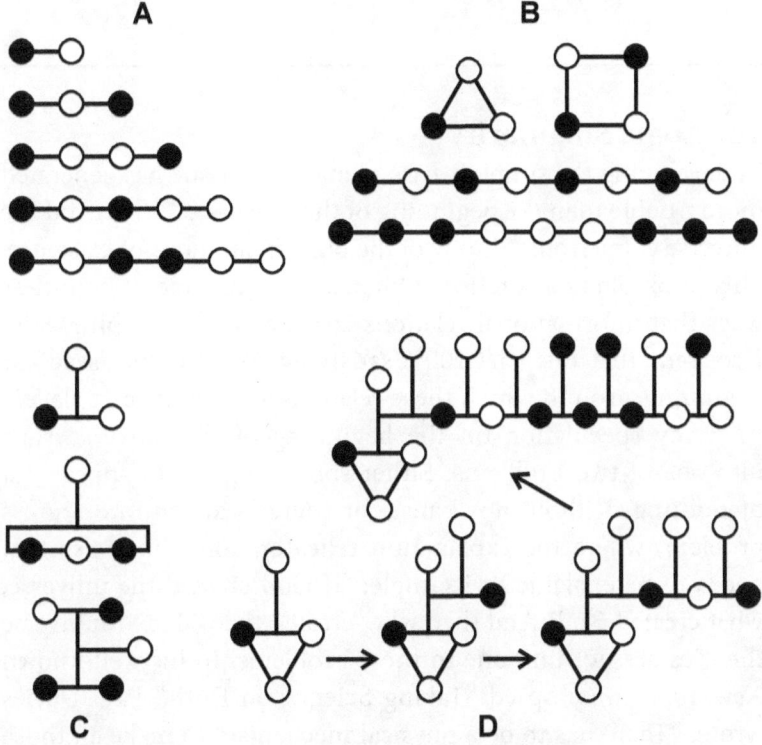

Fig. 6. Various ways that informational relations can combine, each of which is a primordial informational system: as a series of relations (**A**), in ring-like or repeating patterns (**B**), relations with other nested relations (**C**), and combinations of all three (**D**) — here, one system is depicted as building in complexity in steps, similar to the way that our universe evolves.

Now, again you might ask: In what sense are these relations informational, and for whom? Well, if you can imagine yourself as a dot, involved in some unspecified relation with an unknown dot, that other dot could be the same color, or it could be different. It could be either; you are uncertain. But by being part of a *specific* relation — for example, the dots are the same color — then that either–or uncertainty becomes resolved. The relation represents information about sameness, and that information is specific or relative to you. After all, if you were the other kind of dot, the relation would be one of difference.

The Logic Singularity

In developing the simplest-case scenario, no issue has generated more trouble than the beginning of the universe. "Where did the universe come from?" is one of the biggest questions of existence. This book offers a solution: Our world is just one of countless ways that informational relations could possibly combine. Still, I concede that this isn't 100% satisfying. What is the "seed" or beginning, around which these relations form in the first place?

Any speculation on the beginning of the universe runs into one of two problems. Either something has to appear out of nothing, without any cause; or there is an infinite-regress problem, where the explanation relies on something else that needs to be explained. (Example: "If God created the universe, what created God? And then what created that? Etc.) Multiverse theories are not immune to these problems. In his well-known *New York Times* op-ed "Taking Science on Faith," Paul Davies wrote, "There has to be a physical mechanism to make all those universes and bestow bylaws on them. This process will require its own laws, or meta-laws. Where do they come from? The problem has simply been shifted up a level from the laws of the universe to the meta-laws of the multiverse." Both theological and scientific explanations suffer from these problems.

Imagine yourself in a world where everything looks identical: Everywhere you look, you see sameness. You're still getting information about the world, but nothing new is happening. But then suddenly, you see something different. Glimpsing that sight would be new information, telling you that there are in fact things different from everything you've been seeing. This is a little like the experiment in Chapter 1, where we waited for a Geiger counter to click. When it finally clicked, we had new information — something *happened*. Ultimately, our it-from-bit universe is made of these happenings.

I have a name for this seemingly unavoidable difficulty: the *logic singularity*. Our universe (and perhaps every possible universe) obeys laws of logic and consistency. Just as a Plinko chip doesn't make discontinuous jumps from one side of the board to the other, or skip across several rows of pegs, so too our universe follows laws of logical order, resulting in consistent cause-and-effect action. But there's one place where this breaks down: at the beginning. No matter what explanation you consider, causality breaks down there — there's no "before" or cause; there's just "after" or effect. That's the logic singularity.

Remember that in the conventional explanation of *stuff*, some 10^{80} fundamental particles spring into being after the Big Bang. The narrative of this sudden appearance of stuff has become mainstream, and most scientists consider that it's simply the truth — but many admit that it's still mysterious. So, even though the simplest-case scenario doesn't eliminate the logic singularity (I suspect no explanation could), at least it suggests simple and gradual beginnings, rather than a sudden beginning in which a staggering number of ontologically extant particles appear spontaneously in the world, in a cosmological blink of the eye. Simpler might be better in this case.

A Lego Universe

One particular configuration of primordial informational relations is the beginning of our universe. It begins with just one bit — a single relation — and it builds from there in one of many possible directions. In our case, the resulting structure has been developing in complexity for a very long time, as we clock-building, time-measuring humans would put it. Like one particular path that a chip might take down a Plinko box, the universe is one particular way for simple binary relations to combine. To attempt yet another analogy: The universe is like a tree made of Lego blocks, each block being one of these informational connections. At the present point in our history, the tree includes a great many clusters of Lego-blocks made of information, which we call biological organisms. The most complex of these organisms or informational clusters are conscious observers — you and I are two of them. The roles we play within the overall structure, and how the clusters relate to each other, are the subject of Chapter 8.

What is the exact configuration of relations at the beginning of our universe? I wish I could say, but I'm not smart enough to figure that out. Computer simulations could be run, though, and perhaps 50 or 100 years from now, we will have a handle on our informational beginnings. We'll know the first few jogs that the cosmic chip needs to take in the Ultimate Plinko Box, in order to head in the direction of a universe like ours, with sentient beings capable of investigating their world, finding matter and complex chemistry, discovering the laws of how everything works together, and reflecting on what it all means.

I hope you're starting to see how a universe such as ours could evolve from very simple to very complex, while still being much less complex than conventional wisdom would have us think. We look out and see a vast, possibly infinite realm of space, with an incredible number of particles arranged in interesting and beautiful patterns. We probe the micro-realm and find molecules, then atoms, then nuclei, then subatomic particles such as electrons and quarks — things that we call fundamental constituents of matter. However, we can't be sure that the next generation of particle accelerators won't discover particles that seem to be even more fundamental. Smashing particles at higher and higher energies may be as endless as calculating the value of π to thousands of digits, looking for an end to the sequence of numbers. It's possible that drilling down at matter in this fashion won't ever reach a fundamental bottom layer of reality.

In the stuff view, quantum behavior is routinely called weird because it doesn't match up with our intuitive picture of atoms and molecules colliding like billiard balls. Meanwhile, the problem of consciousness is baffling; we aren't close to understanding how things like thoughts and emotions and pleasure can emerge from nerve impulses. As I've mentioned, some philosophers and even a few scientists believe that we will never figure out reality by dividing the world into smaller and smaller pieces, that this reductionist approach was off-track all along.

This book offers an alternative. Chapter 1 argued that the stuff of the world emerges from an underlying system of information, rather than information being derived from stuff. Chapter 2 suggested that information appears in the world progressively, and therefore that the world did not begin with a mind-boggling number of particles of matter, each one "imprinted" with information, as if assigned on Day One by a Creator. Chapter 3 reviewed how information is always relative to some observer; there seems to be no information that's absolute or observer-independent, but rather, information always comes in the form of relations. In Chapter 4 we saw that information reduces uncertainty, and in Chapter 5, that information can compound with other information to reduce uncertainty further still — always relative to some observer. We saw a way to generalize quantum mechanics, specifically Carlo Rovelli's relational interpretation, so that it is entirely about information in its purest, consistently relational form. And, this chapter suggests that there are many ways for informational relations to combine, and that Max Tegmark's Level IV multiverse and the many-worlds interpretation are two ways of looking at this multitude of possible combinations, where Plinko-like probability is an overriding principle that governs how information appears in the world.

By proposing that the potential combinations of relations are like the potential pathways that a chip might take through a Plinko box, the simplest-case scenario provides a way for there to exist many potential universes, just one of which is ours — fine-tuned though it appears to be — without being weighed down by the requirement of trillions upon trillions of material universes coexisting, each containing trillions upon trillions of particles, whose origins still need to be explained. Critics sometimes describe the multiverse and many-worlds interpretation as profligate or absurd under the stuff picture. The simplest-case scenario addresses these ontological objections: The universe is *not* a vast expanse of galaxy clusters, sharply defined down to every last dust grain and speck of supernova-effluvia, extending even well past our own Hubble bubble of observable space and maybe even to infinity. I hope that by the end of this book, even one extremely large and complex universe — with vast amounts of space, matter, and energy — will seem profligate and absurd to you.

In the next chapter, we'll look at why the distant past appears to be nearly as informationally rich as the universe today, why we find evidence of a Big Bang 13.8 billion years ago, and why this evidence presents itself to us, even though the universe's informational beginning could be very different and much simpler. After that, we'll look at how all biological life is really just *one* single system that observes the same world with the same course of events. I'll give you a hint: Both of them have to do with informational relations. Are you surprised?

Things to Remember From Chapter 6

• Multiple universes (collectively known as the *multiverse*) have been proposed to explain why one universe, our own, appears to be fine-tuned to support matter and life. This is analogous to the many threads of biological life across the eons, one of which has led to you.

• Multiple universes can be visualized as the multiple paths that a ball can take through a Plinko box. In this analogy, our own universe is one such path; alternate universes are other paths.

• The path that a ball has taken through a Plinko box constrains what can happen to the ball next. Similarly, the history of our universe, from its beginning to the present, constrains what events can happen in our universe next.

• In the same analogy, the ball's evolving path through rows of pegs is like the accumulation of bits in an informational universe, starting with a single bit and building from there.

• The universe is one possible arrangement of bits, out of the many possible arrangements that these fundamental, irreducible informational relations can form.

CHAPTER 7

INFORMATIONAL RELATIONS ACROSS TIME

Quick — think of an informational relation that you're involved in. That is to say, think of something, some object that you have information on, and think of what that information is.

I'm guessing that you chose something other than yourself, some object "out there" some distance away, and you described a property that the object has right now, in the sense that you could go over and measure it to confirm that information. For example: "That chair — it's green." Or, "The tree in the yard — it has eight apricots on it." Or, "My car — it's a 1993 Chrysler LeBaron."

These statements represent informational relations across space: You are separated from the object, and that separation is along spatial dimensions, a distance that can be measured in feet, or miles, or whatever distance unit you choose. No matter what the object is or how far away it is, we tend to think that we're describing the object as it is right now or "at this very moment" — especially if we're looking directly at it.

Of course, light travels at a finite speed, so we're never really looking at an object as it is right now. For example, the Moon is a little over one light-second away, so when we look at the Moon, we're seeing it as it was about 1.2 seconds ago. Also, if the Sun exploded, we wouldn't know about it for eight minutes. Establishing an informational relation across space *necessarily* means that we are establishing one across time, as well.

This chapter will look at informational relations that extend across stretches of time. Despite seeming obvious, the idea of an informational relation across time is one I've never come across anywhere, either in popular science writing or the physics literature. I believe this is because people disregard, forget, or haven't really come to terms with the intrinsically relational nature of information (Chapter 3). We routinely think of information as an absolute quantity that's "about something," some absolute object. In measuring the velocity of a speeding rocket, or the difference in height between two trees, we acknowledge that these are not only relations (where one value is compared to another, in these cases velocity or height), but they are also relations across space, where the things being compared are separated by distance. But similar kinds of relations can reach across time: This is what happens when you, as an observer, have information on the properties of something as they stood at some previous point in history. A velocity value is a kind of relation across time, since it compares positions at different times, and it's quantifying the rate of the change in that position.

With the help of a Stephen Hawking paper, I will argue that relations across time are the key to a deep understanding of the universe in the distant past, particularly at the Big Bang.

Examples of Relations Across Time

How much did you weigh exactly one year ago? If you know the answer to that question, then that's an example of an informational relation across time: Here you are today, and you have information on a property of yourself from a year ago. You might say something like, "I *now* weigh three-quarters of what I weighed *one year ago*." This sentence combines the number three-quarters (or 0.75) with several contextual constraints, to make a sharp, meaningful statement about a relation across time. Take away any of the elements, and the statement is less sharp and less meaningful, and therefore less informational. It leaves your audience with some uncertainty about what you're trying to say.

How loud was the roar of a tyrannosaurus? We don't know; it may not have roared at all, except in the movies. As such, we have no information, and the answer is uncertain — but for now, let's assume that it roared. If we could get in a time machine with a decibel meter and go back several dozen million years, conceivably we could hide behind a rock like on *Land of the Lost*, and measure the dinosaur's roar. It's difficult to argue that the dinosaurs, if they did in fact make sounds, could not have been measured for this property. So, we can make an estimate — 100 decibels, perhaps? Zero decibels is defined as the threshold of human hearing, so this relation compares the physiological properties of two species that never came close to co-existing, except on *The Flintstones*. Even though we can only make a ballpark estimate, the actual relation, in units of decibels,[1] could be described in principle. The average loudness of a tyrannosaurus and a sound at the average threshold of human hearing is some particular number. We may not know what that number is, but it is a number regardless, in the same way that the first person to fire a shot in the American Revolution was a real person, despite his identity being lost to history.

Continuing on the dinosaur theme: How *tall* was a living tyrannosaurus? Here we don't have to guess; we have some information that would lead us to a fairly accurate estimate. By finding numerous bones and figuring out how they connect, paleontologists have assembled entire tyrannosaurus skeletons, which can be seen standing in natural-history museums. We may not know exactly how tall a *living* T-rex was, but the information provided by the skeleton reduces our uncertainty of that figure, down to, let's say, 5%. Each bone provides mutual contextual information, much like the photons in the one-photon-at-a-time double-slit experiment (page 100).

---

1 The decibel is not like most units. If we measure the levels of two sound sources in units of acoustic power, and we then divide the numbers, we are left with a unitless number — a ratio. The decibel "unit" is actually just a label saying that we have taken the logarithm of that ratio (which gives us a number of "bels") and multiplied by ten.

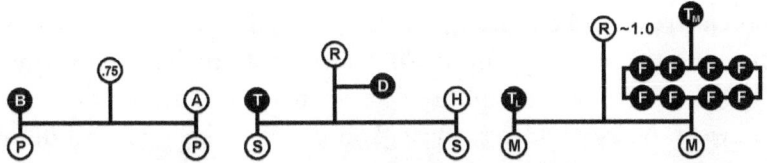

Fig. 1. Examples of relations across time: Left, a ratio (.75) of weight before (**B**) vs. after (**A**), as measured for example in pounds (**P**); the horizontal line represents the relation across time. Middle, loudness in decibels (**D**) of a tyrannosaurus roar (**T**) vs. the threshold of human hearing (**H**), both in arbitrary units of acoustic power (**S**), as determined by the bare ratio between them (**R**). Right, T-rex fossils (**F**) combine to produce the height of a museum T-rex (**T$_M$**), arbitrarily expressed in meters (**M**), where the ratio (**R**) compared to a living T-rex (**T$_L$**) is assumed to approach 1.0.

The point of these examples is to demonstrate not only that informational relations can extend across time, but that such information works the same as all information, whether it involves dinosaur bones from millions of years ago, or photons arriving over a period of hours, or telegraph signals from 100 miles away, or words in a sentence spoken in the here and now. The same principles apply uniformly to wildly diverse aspects of existence, from the early universe to the origin of life to consciousness to language. Information is what all of these things have in common, which is why a deeply informational perspective allows us to make truly coherent sense of the world.

Two Types of Descriptions
Throughout this book, I've used the word "description," perhaps without an adequate definition. Usually, when we think of a description, it's a sentence or a paragraph that paints a kind of picture. You can describe a tree, or a thought, or a galaxy, or a theory in words, so we usually think of descriptions as statements made by humans using language. But physicists and mathematicians use the word in broader ways. "Description" can refer more generally to an abstract relationship among things in nature. For example, the number π describes the relation between a circle's circumference and its diameter. Or, the spin of

an electron can be described as up or "in that direction" relative to some axis, even by a measuring device with no human present. An equation is a description; $E = mc^2$ describes the relationship between energy and rest mass. You could make a general definition that a description is some packet of information, perhaps combining several little sub-packets of information, that can reduce uncertainty for some party in some respect. So, the sentence, "I weigh 0.75 times as much as I weighed one year ago," describes your weight now in terms of your weight last year. $E = mc^2$ describes a quantity of energy in terms of a quantity of rest mass and the speed of light. Both of the descriptions offer the capability to completely reduce uncertainty about something: Armed with those descriptions, you or anyone can determine my exact weight today if I tell you my weight one year ago, or you can determine the amount of energy equivalent to 1.2 grams of matter.[2]

Descriptions of an event can vary, depending on how much contextual information is built into the description. We have seen that a description can be very vague, reducing uncertainty very little; but if the description includes built-in context sharply constraining what is being expressed, then the description can reduce uncertainty by a lot. Think again back to the experiment at the beginning of Chapter 1. A description of the event that contains very little contextual information would be, "I heard a click," or even more vague, "Something happened." These statements may be accurate, but they don't express much of anything or reduce uncertainty in any specific way. A description of the same event but containing a lot of context would be, "An alpha particle was detected coming from a sample of uranium in the physics lab last Thursday at 12:01 PM." This description contains a what, a when, and a where — and if we bring to the situation additional

---

2 The equation can only be meaningful if we know what c, the speed of light, stands for. It is typical that a description requires the receiving party to possess certain information in order to understand or decode the description. This is the basis of all language (discussed in Chapter 9); recall also the SOS story from Chapter 5.

legacy information or knowledge about radioactivity, we can also infer a how or a why (it occurred due to the process of nuclear decay, which is a consequence of the electroweak interaction, etc.). Even this simple example is a relation across time, since you are describing something that happened in your past. The phrase "last Thursday at 12:01 PM" specifies the length of the time interval across which the relation spans.

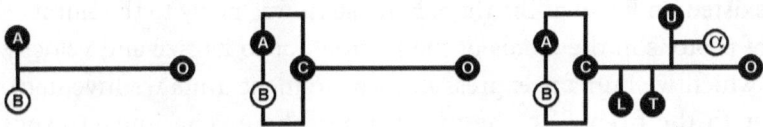

Fig. 2. Descriptions of the same event, with various amounts of context: Left, *something* happened (or happens): **B** is different from **A** for observer **O**. Middle, the event is a click (**C**). Right, the click indicates that an alpha particle (α) was detected at a place and time, because we know that uranium (**U**) was present in the physics lab (**L**) at 12:01 PM Thursday (**T**).

Now, let's think of a situation, an actual historic event, in which contextual information arrives gradually. About 66 million years ago, a massive chunk of rock some six miles in diameter, rich in the element iridium, smashed into the edge of the Yucatan Peninsula, near the present-day town of Chicxulub, Mexico, likely altering the global climate for years and causing a mass extinction of three-quarters of the species of life on Earth. Or at least, that's the description of the event that we have today, pieced together by scientists. They discovered a sudden discontinuity in the fossil record, as well as high levels of iridium in the geological layer known as the K–Pg boundary, and then the Chicxulub impact crater was discovered, and so forth. Surely, when we modern humans look back at what happened, we see a chunk of matter made of atoms, many of which had 77 protons, (the atomic signature of iridium), creating a highly energetic, fiery impact that extended well into the upper atmosphere. We have dramatic artist renderings and CGI recreations of the event, which are probably fairly accurate, in the sense that they're close to what we would have seen had we been watching with our own eyes. That is a picture we arrive at today, because we are able to gather geological, paleontological, chemical, and physical information in so many diverse ways. For example, we have put rock samples into a chromatograph and found a surprisingly high proportion of iridium atoms, each of which we know from previous research contains exactly 77 protons. But that highly detailed narrative isn't the description or the information that existed on Earth at the time. No system was privy to the number of protons in the atoms of the asteroid, or to its size and velocity (which we humans express in measurement units we invented), or to the fireball in the upper atmosphere. The impact event was far more informationally deprived, relative to the simpler systems that were present when the event occurred.

This is significant, because if things like matter, and space and time and the electromagnetic field all emerge out of a bedrock of information, as John Wheeler believed (Chapter 1), and if information arrives in the world gradually, wiki-world-style (Chapter 2), and if information sharpens other information (Chapter 5), then we now have an opportunity to see how complexity in the world can evolve from a point of extreme simplicity. In the stuff picture, primordial beginnings — particles coming out of the Big Bang, or organic molecules coming together to create the first organism capable of metabolism and reproduction — are assumed to happen rather suddenly and definitely. This has led to problems involving apparent fine-tuning, as well as difficulty in replicating the appearance of a new biological system by starting with off-the-shelf organic chemicals. But in the informational picture, these specific, definite processes actually occur with much uncertainty. Only later, when the events become sharpened and enriched relative to organisms like us that have registered and stored large quantities of information, do all of the details emerge — and these details seem to imply that the processes were finely tuned and very unlikely, perhaps to the point of appearing to be miraculous.

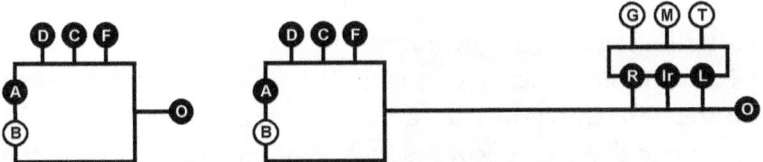

Fig. 3. Left, 66 million years ago something happened, which was likely felt by contemporaneous observers (**O**) as darkness (**D**), then cold (**C**), then loss of food (**F**). Right, we modern observers interpret the same event with greater informational richness. Knowing about the geological record (**R**), iridium (**Ir**), and the impact location (**L**) tells us about global atmospheric effects (**G**), the object's mass (**M**), and the length of the relation across time (**T**).

In order to see how that happens, though, we need to understand that there's a distinction between modern, present-day descriptions of past events, versus *contemporaneous* descriptions that occur within the time frame in which the events are happening. A modern description is how we see the world right now, taking into account all past discoveries about the world, as well as all logical truths, mathematical relations, and scientific theories that we have discovered or developed. The modern description of the world is very detailed; it doesn't leave much to the imagination. When we look up at the stars, for example, there isn't a whole lot of uncertainty about what we're looking at, down to the atomic level with our understanding of the nuclear fusion mechanism in their cores. That was not the case when Paleolithic people looked up at the stars. Their contemporaneous description was different, and much lower in information, than our modern description of the same stars.

A contemporaneous description represents a minimum of information; it's like a baseline or a starting point. An observer might have to wait awhile for other information to come in that clarifies the nature of the thing or event being observed. Before that happens, they have a contemporaneous description. This is the case regardless of what kind of observer you are. Ten billion years ago, as far as we know, there were no systems that registered, processed, and stored information in the way that living things do. So, even though we humans can clearly imagine the universe ten billion years ago, including specific supernovae creating specific atoms of heavy elements that would one day become precious-metal deposits on Earth, informationally we have to realize that that's a *modern* description of the distant past. In the contemporaneous description of the universe ten billion years ago, there were no such things — there was only the *potential* for such things to be so clearly described and understood. If we could place ourselves on the scene ten billion years ago, the modern description is probably almost exactly what we'd observe. But again, as far as we know, there were no such complex observers on the scene at the time. Uncertainty reigned, as if the universe were in many different configurations at once, and yet in no one configuration in particular. Informationally speaking, the universe was much simpler.

The two kinds of descriptions across time, contemporaneous and modern, are related to the themes from the previous chapter. When you think about the long, unbroken chain of all of your ancestors surviving to reproduction age and then successfully reproducing, you're giving that course of events a modern description. It takes into account information about the history of the Earth, the evolution of life, that asteroid impact 66 million years ago, genetics and reproduction, and the survival of every ancestor. That's much more information than an organism many millions of years ago was privy to. The modern description of our ancestry is astonishing, because thanks to the work of paleontologists and archeologists, we know so much about what our ancestors endured and the steep odds they overcame. This modern picture of the world and its history continues to get progressively clearer. But, life on Earth 800 million years ago had no idea where it came from or where it was going. This makes sense, because as observers, those organisms just did not register as much information as much later, more complex life forms like you and I do. Simpler observers correspond to a contemporaneously simpler universe; more complex observers correspond to a modern, more complex universe.

The distinction between modern and contemporaneous descriptions is subtle, but it's critical to understanding the early universe. Every event in the past can be described either in a modern fashion, as imagined by us modern humans with all of our technological tools, knowledge of the Standard Model, the laws of physics, and so on. Or, it can be described in a contemporaneous fashion, relative to the observers that were around at the time of the event. The description of the Big Bang that you see on science shows, for example, is a modern description of the Big Bang. It's what would be seen by people, if we could travel back in time, taking all of our information and scientific equipment with us, and observe the event with our own eyes and instruments. The contemporaneous description of the Big Bang, though, is different; it's more uncertain, and informationally it is far simpler. Contemporaneous descriptions point to much simpler beginnings.

To be clear: The idea of a contemporaneous description does not exist in mainstream science. In the conventional "stuff" view of the nature of things, it is assumed that the way modern science describes a molecule is simply the correct description, at any point in time, since many of its material qualities have been defined, book-world-style, all along. This is where the informational view of the universe offers something different: Since information is relative to the systems that register and use it, the nature of things billions of years ago winds up being far simpler, within that timeframe, than we consider today. The complexity of the universe that we marvel at — which makes us invent gods to understand how the richness of existence could possibly have come into being — is mostly a modern artifact, a function of our human and technological ability to perceive the world at very high degrees of precision. We smash particles at enormously high energy levels, hoping this will lead to a "theory of everything." But if instead we concentrate on how information interacts with other information, *even across time*, perhaps we can grasp a more explanatory big picture of how everything came to be.

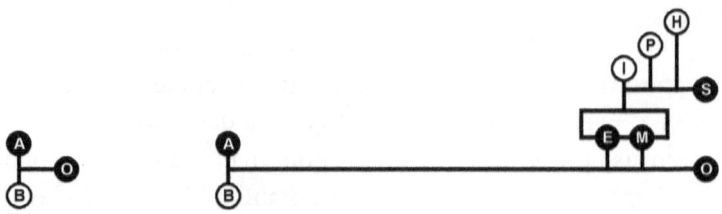

Fig. 4. Left, the very simple beginning of the universe in a contemporaneous sense: Something happened, relative to any observer **O** that may have been present. Right, the beginning of the universe in the modern sense: Measurements of cosmic expansion (**E**) and the microwave background (**M**) tell us about cosmic inflation (**I**). Interpreted within the context of scientific knowledge (**S**), that tells us about the differentiation of the fundamental particles (**P**), the formation of hydrogen atoms (**H**), and so forth.

More Than "Just Philosophy"

Perhaps this is sounding to you like a philosophical or meta-physical exercise — that we can describe the early universe any way we like, but our descriptions have no bearing on what the universe was really like shortly after the Big Bang. But we are suspending our unsupported belief in the assumption that every subatomic particle that came out of the Big Bang had a definite and specific ontology, each particle "pre-printed" with its absolute, characteristic information such as mass and charge, and that it continued on through space and time and countless interactions. We are considering, instead, that information at the subatomic degree of precision did not become a feature of the world until billions of years later, as organisms evolved and built tools that were capable of probing nature to such deep informational levels. If information is fundamental, and if stuff emerges from information, then photons and electrons are not an intellectually honest way to describe the early universe in a contemporaneous sense. The sharpness of individually spec-ified photons and electrons, that degree of complexity in the world, is something that did not emerge until an informationally rich period of history much later. This greatly changes what we can say about how the world "really" was in the distant past.

We've seen that all information is relative to the observer. Also, the *value* of information — its capacity to reduce an observer's uncertainty — depends upon the older information that the observer brings to the table. The more old information that the observer brings along, the more complex (we often say meaningful) the new information may become. The old information constrains the new information: As in the case of the telegraph operator's SOS signal, legacy information, including the knowledge that leads to conscious interpretation, is what gives complexity to information that we receive from the world. It therefore becomes meaningless (quite literally!) to describe the early universe as having any contemporaneous complexity at all, when there were no complex observers around to have this description. If the fundamental nature of things is informational, and if observers in the distant past are simple, then the early universe necessarily must also be simple in a contemporaneous sense. It turns out that this argument is consistent with a theory from the former Lucasian Chair at Cambridge.

Hawking's Top-Down Cosmology

The previous chapter discussed the apparent fine-tuning of the universe. One of the reasons why fine-tuning is a problem is that for the universe to be fine-tuned today, its initial conditions, at the Big Bang, needed to be balanced exactly just so in order for its seemingly fine-tuned modern properties to be in place. Indeed, that's partly why the theory of cosmic inflation has been embraced: It could produce a multiverse (pages 144–148) that would include some universes with the exact initial conditions that would lead to a finely tuned mature universe such as ours.

This could be called a "bottom-up" approach to cosmology: The goal is to find the precise initial conditions (at the bottom — physicists like to imagine time flowing upward) that would produce the universe that we currently see (at the top). In other words, we know the present conditions, so we try to figure out what precisely tuned conditions at the Big Bang would produce those conditions today. It's kind of like the Plinko analogy, if you imagine the box upside-down, with the chip falling upward. Observing where the chip wound up at the end of its path, we try to determine *precisely* how it would need to be released into the box in order to take the path that it did. This is a primary focus of the quest for the "theory of everything" in physics: It's the search for one theory, one master equation, that would predict the precise initial conditions of the universe and therefore the universe's present conditions as well as all of the laws of nature. That's the bottom-up approach to cosmology. To say that this approach has been frustrating would be an understatement; read Lee Smolin's *The Trouble With Physics* for an expert dissection.

In the mid 2000s, Stephen Hawking, along with Thomas Hertog of CERN and with inspiration from the physicist James Hartle, proposed an alternative: What if, instead of pursuing the intuitive bottom-up approach, we try the exact opposite, a *top-down* approach to cosmology? The authors suggest that by looking for a distinct, definite, and individual set of initial conditions for our universe that would lead to the present conditions, we are assuming that the universe changes according to classical mechanics — that it's an ordinary object which is in one particular, definite state at any given time, like a metal coin physically tumbling through the air. While such a picture may be attractive to the intuition, it is deeply problematic. "A bottom-up approach to cosmology either requires one to postulate an initial state of the universe that is carefully fine-tuned — as if prescribed by an outside agency — or it requires one to invoke the notion of eternal inflation," Hawking and Hertog write. "Here we put forward a different approach ... based not on the classical idea of a single history for the universe but on the quantum sum over histories."

"Sum over histories" is the mathematical method that accounts for the motion and change of a quantum system over time. It says that a quantum system doesn't take a single, classical trajectory to get from one place to another, like a bullet; instead, the object takes literally every possible path, at least in the case that the observer doesn't have information on which path it is taking. This is closely related to the observation that photons in the double-slit experiment (pages 37–41) act like waves that interfere with each other, i.e., each photon passes through both slits at once, or simultaneously passes through and is reflected by a half-silvered mirror (page 37). If the universe is a quantum system — and it would have to be, if it is governed by quantum mechanics, as many believe — then, Hawking and Hertog argue, we cannot treat the universe as a classical object with a single defined history. Instead, when asking, "Why is the universe the way it is?" we must consider *all* possible histories. Put simply, the universe begins in every possible manner that a universe could begin, even though there is only one history that we observe today. This singular observed history suggests that the beginning must also have been singular. But Hawking and Hertog say that this assumption is wrong.

In terms of the universe, this is a way to explain fine-tuning with a multiverse theory of sorts, without actually requiring an ontic multiverse, where for example many trillions of variations on our own universe materially exist but are too far away to observe (Level I and II multiverses). Instead, there is a landscape of *potential* universe histories, all beginning in an indistinguishable manner, and then branching off from each other. That idea is consistent with the proposal of a Level III or IV multiverse, and with multiple universes arising out of possibilities (Chapter 6). The multiverse is like a Plinko box, our own universe being one of the many paths that can be taken.

Top-down cosmology ties in with contemporaneous and modern descriptions, as well. The observable state of the universe, at later times such as the 21st century, is a modern description; it's very rich and detailed, all of those details suggesting a unique history back to the very beginning. The very early universe as described in conventional bottom-up cosmology, with (one

assumes) finely tuned initial conditions, is also a modern description: The only way we can extrapolate that unique set of
initial conditions is to consider everything we know about
the state of the universe today, and work backward. That makes
it a modern description. But Hawking & Hertog's description
of the universe starting in every potential manner — having
all possible initial conditions at once — is a good example of a
contemporaneous description. When the universe is very young,
there is no unique set of conditions; the description is uncertain
and undefined, and the universe is extremely low in information
content, relative to any contemporaneous observer. But there
exists the potential for the system to acquire more complex configurations of relations, and this can happen in various ways (as
in **Fig. 5**). This is how the system can split into branches that
are distinguished from one another: With information building
upon information, eventually complex subsystems, which we
call observers, appear within some branches. These observers
register the new information appearing within their branch, and
this information accumulates as legacy information, resulting in
increasingly complex descriptions of their world. That's where
we come in: Complex observing subsystems are what we are, and
registering and digesting information is what we do.

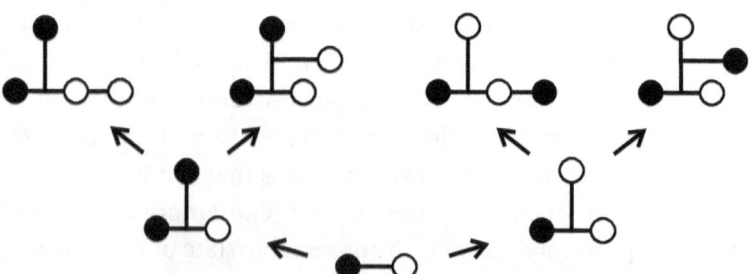

Fig. 5. Bottom center, the informational beginning of all possible universes,
with other universes branching off as different possible configurations
(not all possibilities are shown). Some branches will develop complex
observers, but an observer will see only its own branch.

Chipping Away at the Universe

In their top-down cosmology paper, Hawking and Hertog don't speculate on how a universe that starts in every possible manner winds up being in an apparently distinct configuration, as we observe. In the physics literature, there is much discussion about *decoherence* (discussed in the Appendix) and something called "environment-induced selection." These are stuff-based explanations for the appearance of wavefunction collapse, the tendency of nature seemingly to snap from uncertain and wave-like to defined and particle-like when a measurement or other interaction occurs. In the stuff view, these mechanisms could explain how our unique universe emerges out of Hawking and Hertog's landscape of configurations. But if we remember John Wheeler's it-from-bit conjecture, that the material world emerges out of a process of observer-participancy, then we see how observer-participancy is a top-down affair all the way: When we study the early universe, we are creating informational relations across time, *from now back to the beginning.*

I've mentioned Paul Davies before. He's the guy who made waves in the *New York Times* by suggesting that physics is becoming troubled by faithful thinking. Davies has received numerous prizes throughout a distinguished career, and he presently has one foot in cosmology and the other in biology and cancer research. As director of the Beyond Center at Arizona State University, Davies is one of the top thinkers in foundational physics and cosmology, and also one of the most willing to go out on a limb and consider radical proposals. For example, he and Beyond colleague Sara Imari Walker[3] wrote a paper suggesting that when asking about the origins of life, we ought to consider information. Living organisms display complex and directed information flows (nerve impulses, biochemical signals, the transcription of RNA

---

3 Walker had a prize-winning essay in the same 2012 competition as my essay on which this book is based. Titled "Is Life Fundamental?" Walker's essay touches on themes relevant to those written about here, particularly in the next chapter.

into proteins, and so on), so Davies and Walker suggest that we should investigate how information might have been harnessed and directed in the earliest life forms. This could shed more light than a strictly chemical approach to the origin of life, where we envision how molecular structures may have come together to form the earliest living systems.

Davies was interviewed by Robert Lawrence Kuhn on the PBS series *Closer to Truth*, in an episode about consciousness. Davies refuted the idea that consciousness points to an intelligent designer of some kind. He suggested that both consciousness *and* fine-tuning could evolve entirely within the universe, through a kind of feedback loop involving the observation process, with no intentional agency guiding the way. Here is what he said:

> In the popular mind, there's this notion that there's a unique history that connects the Big Bang, the origin of the universe, with the present state of the universe. Quantum physics says that's just a load of baloney — that there's an infinite number of histories. They're all folded in together, and if you know nothing at all about the past of the universe, you must take all of these histories. And when we make observations, what we're doing is chipping away at these histories and removing some of them. We're culling them. And in principle, if we could fill the entire universe with observations, we would then home in on something like a unique history. So, the act of observation, in part, resolves something about the histories of the universe.
>
> The laws start out unfocused and fuzzy, [but] eventually there's life and observers, that link back, just like in quantum mechanics, back in time, through making their observations, and help sharpen those laws in a way that's self-consistent with their own existence. So here we have a universe that has an explanation within itself: The observers that arise, play a part in selecting the very laws that lead to the emergence of observers in the first place.

You have to have this. If we're trying to explain why does the universe exist in its present form, and in particular why does it contain life and observers, obviously those life and observers have to be relevant to the laws that give rise to them. Because there's no other way you can have an explanation for the universe from entirely within it.

We have seen that there is a connection between making observations, acquiring information, and reducing uncertainty. Davies says that by making observations — reducing our uncertainty through acquiring information — we are chipping away at the alternate potential histories of the universe, or culling them. The universe starts out in every manner possible, as in top-down cosmology; from there, potential histories are gradually removed from the equation — specifically, those that conflict with information that appears to observers. As a result, the one history that's observed gets more and more specific, more and more informationally rich, like the pathway of a chip through a Plinko box, where one more bit is specified each time the chip jogs either left or right.

Davies uses the word "sharpen": Acquiring information sharpens the universe's laws, which also consist of information. "Link back in time" alludes to informational relations across time. "If we could fill the entire universe with observations" hints of having a complete, modern description of every particle everywhere, up to the limits of the uncertainty principle. But since we aren't able to fill the universe with observations, Davies' statement suggests that every location that has not been touched by observer-participancy remains unfocused and fuzzy. That means high uncertainty and low information, relative to the observers in question — *us*. We routinely assume that for billions of light-years in all directions, the universe is packed with information about every last particle, but Davies' quote suggests that this assumption is false. Instead, the universe is far simpler than we think!

Distant History as Backstory

When cosmologists talk about the finer points of the Big Bang, such as the physical processes that occurred in the first fraction of a second after the event, they aren't just guessing. These details are based on observable data. So when Stephen Hawking and others conjecture that the universe begins with every possible history at once, that seems to contradict observations, which suggest one unique history. We therefore have to figure out why we would get data pointing to one unique history, if the actual, contemporaneous history isn't unique.

I claim that the observed unique history is *backstory* — it's a logical consequence of us observers already having received other information about the universe. To borrow Paul Davies' words, the universe starts out unfocused and fuzzy, with all potential histories folded in together. However, when we do cosmological research on the early universe, we must take into account that we have already chipped away at those histories and removed some of them. All new information we receive about the early universe is constrained by that old information. The universe being a logically consistent place, no new information appears that contradicts information we already have. So it makes sense that we observe an early universe with conditions that would lead to the present universe. Out of an infinite number of histories folded in together, emerges the appearance of *one* unique beginning — the backstory for the unique world that we observe today.

When we look at a distant object through a telescope, it's said that we are looking back in time. If a quasar is ten billion light-years away, we say that we are observing the quasar as it was ten billion years ago. It is very tempting, immersed as we have been in the stuff picture all our lives, to imagine that those photons going *splat* on our retina zoomed across billions of light-years of space, all the while carrying with them information about the quasar they originated from, each photon wavepacket imprinted with a unique direction/momentum, wavelength, etc., book-world-style. As conscious observers who see things as being made of stuff, we very naturally interpret quasar light as energy–stuff coming from a giant collection of matter–stuff a great distance away. And, like the experiment at the beginning of Chapter 1, we have every reason to believe this interpretation: We have studied the heavens for thousands of years, and we have developed an advanced body of astronomical theory. We also know the speed of light, and we can estimate large distances, even in the billions of light-years. Therefore, when we register a photon from a distant quasar, we naturally decide that the photon must have been emitted by the quasar billions of years ago, and that it has been racing through space at the speed of light for all of that time,[4] until finally, *splat* — the end of its long journey. However, recalling Chapter 1, this interpretation is a philosophical stance. If the photon had indeed been out there ontologically zooming through space when Julius Caesar was alive on Earth, we could not determine this. No experiment, certainly not in any kind of practical sense (and likely not even in principle), could ever settle that question for sure.

In the case that information works like a wiki-world, then information about the observed state of a quasar is appearing in the world, for the first time, *when it seems to go splat on your retina* (or on a photographic plate, digital sensor, etc.). The story about the photon zooming through space for billions of years before finally hitting something is just that: a story. This is what Wheeler meant when he wrote, "any talk of the photon 'existing' during the intermediate period [before detection] is only a blown-up version of the raw fact, a count." It is a stuff-based narrative, built around our other stuff-based narratives of the universe.

---

4 Even in the conventional physics of stuff, it's problematic to say that a photon races through space. Using the active verb "races" infers that the photon is doing the racing, and "through space" implies that space is whizzing past the photon, the way a fish swims through water. But according to special relativity, for something moving at the speed of light, time and space both contract to a point. So, even in the stuff view, photons can't really *do* anything in time, or race *through* anything. Things can only get in their way.

When we peer out into deep space and look back in time several billion years, we're effectively putting our modern selves — including all of the information that we've gathered to date — into the scene from afar. This is why we are able to create a modern description of what we're seeing, based on the observations of photons appearing in our world at that moment. The information provided by those photons is constrained by older information. For example, we approximate the distance to a far-off object by the red-shifting of its spectra, the stretching of light across great distances of expanding space. That could not be accomplished without knowing the unshifted spectra of elements such as hydrogen. The standard brightness of a certain kind of supernova goes into the equation as well. Then, combining the distance with the known speed of light lets us approximate how far back in time we're looking — the length of the relation across time. These are all aspects of the sharp modern description we can provide, by being complex observers that have collected and processed all of this information.

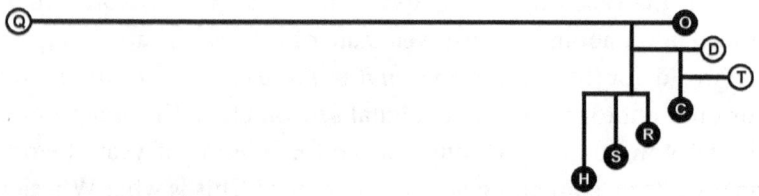

Fig. 6. To observe a very distant quasar **Q** is to establish relations across both space and time. Knowledge such as the spectrum of hydrogen (**H**), the brightness of a Type IA supernova (**S**), and the quasar's redshift (**R**) allows us to approximate the distance (**D**). The relation between that distance and the speed of light (**C**) tells us the length of the relation across time (**T**).

These same principles apply to the very beginning of the universe as well. According to observations, the universe is expanding; almost everything out there is red-shifted and is therefore moving away from us. When scientists extrapolated the expansion backward in time, they arrived at an age of the universe. And, if the universe began with a sudden expansion, then a blast of electromagnetic waves should have been left behind, like echoes. By today those waves should have stretched and cooled to a specific frequency and temperature. So, when it was discovered completely by accident that a buzz of microwaves with these properties permeates space in all directions, it was pretty much irrefutable evidence for the Big Bang — that all the matter and energy of the universe came into being about 13.8 billion years ago. Of course, that's a description in terms of stuff, or if you like, it's a modern description that stretches across a very long expanse of time. It's the ultimate informational relation across time.

The *cosmic microwave background* or CMB, as it came to be known, is an observation consistent with the laws of physics and other information we have. By the time the CMB was discovered, information had appeared in the world, such as observation of the expansion of the universe, that would logically constrain any new information appearing. One such constraint was a heightened potential to observe a particular wavelength of microwave photons buzzing around the Cosmos. We shouldn't be surprised by the discovery; new information being compatible with old information, such as previous observations, is just the consequence of living in a world that's logically consistent. It's no different from measuring the spin of a particle and finding that the measurement is consistent with previous measurements. Or, it's like discovering a quasar at a distinct place in the sky, and then going back the next night and seeing it again in the same spot. If it weren't the case that all present observations are constrained by past observations, the world would not have a logical order, and we wouldn't be able to state any laws of nature.

The photons of the cosmic microwave background were not ontologically born 13.8 billion years ago, in a Big Bang that occurred in a unique configuration out of fixed and apparently finely tuned initial conditions. Instead, the CMB photons represent the appearance of *new* information in the world. And, their measurable properties are constrained by old information in the world. When we measure their wavelength, we should not be surprised to find they have the properties that have been predicted. It's like knowing that Alice owes Bob $100, and expecting to find (and then subsequently finding) an I.O.U for the exact amount.

Does all of this mean I'm saying that a Big Bang didn't happen? Not at all. However, *the* Big Bang — the account you see on science shows, with the breakdown of what happened when the universe was various fractions of a second old — is a decidedly modern description of that beginning. *That* Big Bang,

The Cosmological "Axis of Evil"

One of the most peculiar scientific discoveries in recent years is so perplexing, so confounding, that it has been dubbed the "axis of evil." The cosmic microwave background is extremely smooth and appears very uniform in temperature, everywhere across the sky. In fact, this uniformity is a main motivation for the theory of cosmic inflation: There's no other explanation for the high degree of smoothness if the universe initially expanded at a more pedestrian rate, because shortly after it started expanding, regions became isolated from each other by distance, and so we should see hotter/denser and colder regions of the CMB — a bit like the differences in temperature on the Earth. They aren't there.

The CMB isn't perfectly uniform, though; some regions are very slightly warmer or colder than others. And, if you map these extremely minor differences and remove all known sources of influence, something astonishing emerges: The pattern of warmer and colder spots lines up to a certain extent with the plane of the Earth's orbit around the Sun, and even with the equinoxes. Mind you, this plane is unrelated to the plane of the Milky Way, whose plane is unrelated to that of any other galaxy; all of the orientations are jumbled, so it makes no sense that the

the condensation of specific kinds of matter and energy, and presently known laws and constants out of seemingly fine-tuned conditions, is what we humans would have seen, had we been around 13.8 billion years ago with all of our present technology to witness and measure the event. Those are the informational relations we would be establishing. But since we can't go back in time and observe the early universe in person, we have to make the observations across vast stretches of space and time. And in doing so, we need to acknowledge that other information has already been received — information that sharpens and enriches any new information that comes along.

CMB would be aligned at all with the Solar System. When first discovered, this alignment was assumed to be a mistake or a pollution of the data. But the finding has repeatedly stood up to increasingly sophisticated measurements, including by the state-of-the-art Planck satellite. According to one calculation, the probability that this would happen by chance alone is less than 1 in 10,000. First fine-tuning, and then this — what's going on?

Lawrence Krauss, a physicist unequalled in his skepticism, writes of the problem: "Is this Copernicus coming back to haunt us? That's crazy. We're looking out at the whole universe. There's no way there should be a correlation of structure with our motion of the earth around the sun ... That would say we are truly the center of the universe." According to the stuff view, perhaps; it makes little sense. But if we allow that the universe is informational, there might be an explanation: Information about the CMB is constrained by other information that exists in the world relative to us local observers. The "axis of evil" might be evidence not so much that we're the center of the universe, but rather, that the universe we see is *exclusive* to us. More on that in the next chapter.

If the world is informational, and evolved in one of a great many potential pathways, like paths through an Ultimate Plinko box, then we are able to see how our universe, indeed any possible universe, may evolve from simple to complex. Universes all start out being similar. The available information is low, and uncertainty high for any observers within, hypothetical or otherwise. And universes evolve from there, in the form of the many splitting pathways of potential informational configurations (as in **Fig. 5** on page 180). In this manner, eventually information-utilizing observers find themselves embedded in one universe that starts out very simply, and then evolves to higher and higher degrees of complexity and sharpness, as judged by the similarly evolving and complexifying observers within. For certain sentient observers, the world is seen as being so specific and complex, they conclude that the universe's initial conditions must have been fantastically fine-tuned. This is not unlike a lottery winner, knowing the steep odds they overcame, asking, "How did I get so lucky?" It's also not unlike a person reflecting upon their thread of ancestry going back three billion years and asking, "How did my lineage survive?" Like telling ourselves that we are blessed and protected by a god, we tell ourselves that our universe began in a fine-tuned manner. However, that's only a result of discovering the universe's rich informational backstory — the antecedent conditions that necessarily would have been in place at a Big Bang of *stuff*, in order to produce the complex modern world we observe and know so well today.

Occam's Return

I've argued that we need to distinguish between contemporaneous and modern descriptions of past events if we are to understand the true nature of the early universe. This allows us to see the history of the universe in more of a bird's-eye view, even though we are trapped in a frog's-eye view. Admittedly, all of this may seem complicated! So, you may ask: Isn't the conventional *stuff* explanation simpler? Doesn't Occam's razor (page 23) therefore suggest that it's the correct explanation?

You have to understand, though, that the devices and analogies I've introduced in this book are only intellectual tools to help us wrap our brains around what's really going on. These tools are necessary, because we've never thought of the universe before in terms of informational relations, and we need assistance in comprehending this very unorthodox way of thinking. Things like contemporaneous descriptions, bird's-eye views, etc., are artificial constructs and play no role in the emergence of reality. I write about them only because they help us to understand. We seem to want to interpret the world as consisting of stuff out there, so our intuition, and even the language we use, make the stuff approach *seem* simpler. I'm sure Copernicus' critics similarly believed that a stationary Earth — "Earth" at the time being the very definition of what it meant to be stationary — was a simpler explanation than one that depicted Earth careening through the heavens. But, like the Copernican way of looking at the motion of the Solar System, it's time for a new picture of reality. The view of persistent particle–waves, fields, and a spacetime background is useful and practical in many ways, but ultimately it's a picture that's far more complicated than it needs to be.

There is only one fundamental constituent of the world: the binary informational relation, or bit. Every observable thing emerges from the ways that this one bit can combine into complex structures of information. I use dozens of devices, diagrams, and tortured analogies to get that idea across. But as for how the universe is actually put together and what it's made of — it literally can't get any simpler than that.

Things to Remember From Chapter 7

• Just as an observer can have information about something across space, an observer can have information across time, regarding an event that happened or an object that existed in the past, including the very distant past.

• In accordance with the principles of informational mechanics (Chapter 5), informational relations across time are sharpened and enriched by contextual information, including knowledge. The accumulation of old information sharpens new information.

• The very distant past, such as the first few moments after the Big Bang, appears complex and richly detailed to us, because any new information regarding such events is sharpened by information that we already have. Additionally, scientific knowledge builds up the context in which we interpret modern experimental observations. This results in a rich, stuff-oriented description of the Big Bang.

• Contextually sharpened information is what makes up the *modern description* of a thing or event in the past. Nearer in time to the event, however — before contextual information has arrived and has been assembled — an observer in that time period has a *contemporaneous description*. This is information about that thing or event, devoid of context.

• This explains how the universe can begin with utter simplicity, while appearing in relation to us modern observers to begin with finely tuned complexity: In the contemporaneous description of the beginning of the universe, it consists of only one bit. A modern description of the early universe, however, is fantastically rich and sharp. It's what we would see if we could travel back in time, taking all of our knowledge, history, and scientific instruments with us.

CHAPTER 8

THE SUPER-OBSERVING
SUPERORGANISM

The time has come to reveal one of the most controversial claims in this book. It's kind of like the revelation purportedly given only to Scientologists who have reached a certain level of the religion, where followers who have paid large sums of money finally learn about the interstellar tyrant Xenu and his Galactic Confederation.

In other words, you weren't ready for this secret until now.

The claim is that, informationally speaking, the universe did *not* begin 13.8 billion years ago, which is the standard, accepted age of the universe in mainstream science. As discussed in the previous chapter, that beginning represents the start of the modern description of the universe, or the start of the informational *backstory* of the universe. It's what we get when we take all observations of the expanding universe, and we extrapolate those observations backward, to find a beginning of that apparent expansion of stuff — and then we describe what happens at that moment in terms of everything we presently know about the stuff-universe of particles, fields, and forces. Equally, it's what we would see if we could freely travel across the eons in a time machine, bringing all of our modern knowledge and scientific instruments with us. If that were possible, we would observe first-hand a birth of the emergent *material* universe of particle–waves and fields and spacetime — the conventional account of the Big Bang.

But, when we approach the beginning of the universe in terms of *information*, and we entertain the fact that the world is fundamentally informational rather than fundamentally material, we realize that 13.8 billion years ago is not the beginning, in information terms. Actually, for most of that history, our universe is contemporaneously empty in an informational sense. Recalling the analogy of the Ultimate Plinko Box (page 152), not even a single row of "pegs" has been traversed, so the universe contains zero information. It is entirely uncertain relative to any observers that may exist in that time frame, which is to say no observers. The possibilities are wide open; the universe can be any universe it wants to be.

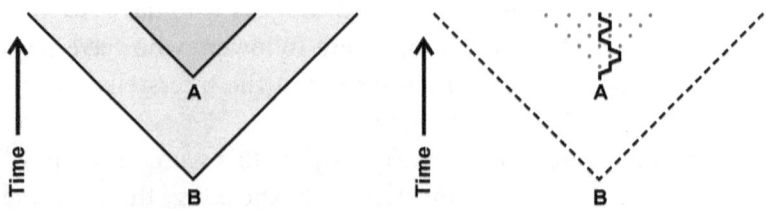

Fig. 1. In the mainstream approach of stuff (left), the universe begins with the Big Bang (**B**) and life appears later (**A**). In the simplest-case scenario (right), the universe begins informationally at **A**, and evidence of a stuff-oriented beginning is discovered much later, through the interpretation of observations by complex observers with advanced scientific knowledge.

So, what happens that sets our particular universe on its own path? What is it specifically that begins to distinguish our branch of the multiverse from other potential branches? Or, in the words of Paul Davies (page 182), if the universe's laws start out unfocused and fuzzy, at what point does the "chipping away" process begin, where alternate potential histories are gradually culled and removed, causing the laws to sharpen?

I claim that if the universe is fundamentally informational, this happens with the process that we call life, or something like it. The event that we modern humans call the onset of life is when durable informational relations first start to appear in the world. *Fundamentally speaking*, our universe does not begin until that moment, when information registration — the kind of informational processing and storage that we see going on in modern-day self-sustaining biological organisms — first occurs. Some ten billion years later than the Big Bang!

Later in the chapter, I will argue that we observers are informationally isolated from any other possible biological-type observers in other star systems or other galaxies. That is why we, the life forms of Earth, or possibly of the local Solar System, are alone in our particular universe. And yet, we are as one single observer, working together across the ages to reveal the same world, bit by bit.

Two Descriptions of the Origin of Life

In the previous chapter, I mentioned that every event in the past can be described two ways: There is a modern description, and there is a contemporaneous description. To review, a modern description is a mainstream-science description: It is the interpretation of an event through many lenses of information we have as conscious observers with a rich history of scientific investigation. The contemporaneous description, however, represents the event stripped of a modern interpretation. It is the raw, informationally minimal description that exists in the same time frame as the event itself, relative to observers in that time frame, whatever those might be. Remember, we are talking about a worldview where information, not material stuff, is the fundamental ingredient of nature (Chapter 1), and also where information appears in the world gradually, as opposed to suddenly at or shortly after the Big Bang (Chapter 2). Mainstream science assumes that modern descriptions are always appropriate — everywhere and at all times — because it assumes that the world is fundamentally made of stuff, and that all of this stuff carries at least some information "written" on it like words in a book. This assumption applies even when the events are in the very distant past, and/or very distant in space. However, we are challenging that assumption, based on the experiments described in Chapter 2, and also because the universe is simpler and more elegant without that assumption.

The origin of life is a good example of an event in the distant past. Suppose we could get into a time machine, taking all of our knowledge and some scientific instruments with us, and go back to a critical moment in the beginning of life. We don't know exactly when or where that might be, because little or no evidence from that event has made it to today. But we can speculate. The time machine might bring us to a hydrothermal vent on the ocean floor, where organic molecules such as amino acids are finding themselves in membrane-bounded bubbles along with alkaline water from the vent. In some of the bubbles, a combination of the amino acids begins to facilitate a swap of electrically charged atoms, or ions, across the membranes, resulting in chemical reactions that release energy. The energy is absorbed by the organic molecules, causing them to rebuild into more complex structures. A few of these structures are capable of harnessing the released energy to replicate themselves, out of similar constituent molecules within the membrane. And in one bubble, everything comes together just right to produce the first minimally functional proto-cell that is capable of growing and dividing. Or something like that.

All of this, of course, is a wild guess. But if it were true, then what I wrote is a modern description: With the help of the time machine, and our scientific knowledge and our instruments, we were able to witness that mysterious moment when lifeless organic molecules began to self-organize into structures that had internal interactions more complex than anything that had happened previously on Earth. It's the moment every paleobiologist or astrobiologist dreams of witnessing. But, now let's abandon the idea of the time machine and the equipment, and try to imagine the same event contemporaneously, according to the simplest-case scenario. What occurred in the time frame of the event itself, informationally relative to this primordial observing system? We might say that the first information-storing molecules were synthesized, but even that description is much too modern. *Something happened*;[1] that is the extent of the contemporaneous description, because from the perspective of what we modern humans would call the newly energized and reconfigured bubble of stuff itself, nothing more than that can be said or known.

The Fine-Tuned Origin of Life

Recall from Chapter 6 that according to many physicists, the parameters and constants of nature create the appearance of the universe being fine-tuned to support the formation of matter and eventually life. This is a motivation for mainstream types of multiverse theories, as well as the theory of eternal inflation: If the bigger picture includes very many universes, all different from one another, then by chance alone, some of them might be fine-tuned for matter and life, as judged by complex observers such as ourselves that eventually evolve within those universes. I argued an alternative view, that the finely balanced initial conditions at the Big Bang are an artifact of modern observation. We observers have been culling the alternate histories, i.e., those that are not "tuned," for billions of years. At this point, we have chipped away virtually all of the non-tuned alternate histories, in a top-down manner, leaving behind only histories that appear to be exceptionally finely tuned.

The origin of life on Earth presents similar problems for astrobiologists and related theorists. We have strong evidence for organisms living on Earth at least as early as 3.7 billion years ago, shortly after the molten crust solidified. This has motivated two theoretical proposals: (1) Biological-type systems appear out of nonliving organic and inorganic matter relatively easily. Life

---

1 It might be better to say that the contemporaneous description is something like 1 (as opposed to 0), where it is understood that 1 represents something happening and 0 represents nothing happening. The choice of representative symbol is arbitrary.

doesn't require extraordinary, highly unlikely, or seemingly miraculous conditions or extremely long periods of time to get going, and therefore, life should not be particularly rare in the universe. Or: (2) Earth was seeded with life from elsewhere very early in its history, perhaps by comet or asteroid bombardment, perhaps even from Mars. Both of these hypotheses are a response to direct observations, in this case the radiometrically dated microbe fossils we've discovered; any theory that is proposed has to be consistent with observations such as these. But the simplest-case scenario offers an alternative. The same informational treatment of Big Bang fine-tuning can be applied to the origin of life. In fact, we might not have a choice.

Similar to the seemingly balanced parameters at the Big Bang, if we could get in that time machine with our microscopes and scientific knowledge to observe the origin of life, we might witness a coming-together of conditions that is *extraordinarily* unlikely. It might even rival or surpass the seemingly razor's-edge balance of the parameters of nature at the Big Bang. It's possible that the sequence of molecular events leading to the first life form might be so unlikely, the probability that it would occur *even twice*, anywhere within the observable universe over billions of years, is vanishingly small.

Now, to any well-informed science enthusiast, this sounds like crazy talk. Why should abiogenesis, the term for the appearance of life out of nonliving molecules, be so incredibly unlikely? It doesn't have to be. But, it *could* be — and in that case, the simplest-case scenario has an explanation. It doesn't matter how astronomically improbable the particular molecular comings-together may have been, as we would describe them in the modern sense. Like the long, unbroken thread of ancestry that has led to you being here today, abiogenesis being incredibly unlikely may be evident only in retrospect, when we apply all of our modern information to bear on the question. It seems to be miraculous to us modern humans, but speaking contemporaneously, *something just happened*. That's all.

Still, abiogenesis is difficult for theorists, and the problems are well discussed in the scientific literature. It's what keeps astrobiologists awake at night. Scientists generally agree that three molecular conditions must be met in order for the first truly biological organism to take hold and lead to future generations of life: (1) self-replication or reproduction, (2) metabolism or the harnessing of energy to perform work at the molecular level, and (3) an enclosing membrane to selectively isolate the system from its environment. Some of these capabilities may have appeared in precursor systems, but at some point, a system had to have all three to function in a truly biological fashion.

In the 21st century, when we have broken apart matter into quarks and leptons and measured the cosmic microwave background to within thousandths of a degree, the fact that there are multiple different abiogenesis ideas — all of them speculative — speaks to the difficulty of this problem. Unlike the Big Bang, though, it may be that we will never acquire enough evidence to nail down this one definitively, so that we may come up with a testable, reproducible modern description of the convergence of materials that produced the first life. But we can try.

The difficulties of abiogenesis are similar to the challenges of universe fine-tuning. When you assume that the world is fundamentally stuff-based, that matter at all times takes specific configurations and the parameters and constants of nature at all times take specific values, then appearances suggest that the Big Bang (and perhaps abiogenesis) occurred in an unlikely, remarkably configured, or carefully coordinated fashion. These astonishing beginnings are partly why people invent gods — and even cosmic-inflation theories. Both are there to fill a hole created by the lack of other explanations. The simplest-case scenario, however, handles both the Big Bang and abiogenesis in the same top-down manner, where their unlikely or fine-tuned qualities are evident only in retrospect. Contemporaneously, i.e., in the time frames of the events themselves, nothing unusual is happening at all.

When we consider the universe's first years under a modern description, we find that complex systems, such as living organisms, couldn't have appeared immediately. For one thing, there were no heavier elements such as carbon, out of which many different kinds of molecules could form. Complex systems such as those seen in life likely couldn't form until billions of years after the Big Bang, by which point in the modern description, there had been sufficient nucleosynthesis of heavier elements, and enough organization and clumping of matter, for certain organic molecules to possibly find themselves on a rocky watery planet, as in the story above.

It's *this* point in the modern description, the onset of life, that's actually the informational beginning of the universe. Before the appearance of the first biological organisms, or other complex, self-regulating systems that register and store information in a robust manner, enduring information just does not happen in any sense at all — so the universe, relative to any extant system, remains contemporaneously "unfocused and fuzzy."

Life, or something like it, seems to be required in order for information to endure or stick in the world. Information that may be associated in the modern description with a configuration of non-living matter, for example, tends to degrade and get lost, as a natural result of entropy increase (see page 66). Things break down in this world; structures like cell membranes don't last very long after an organism has died, and DNA degrades over time as well. But life has evolved mechanisms to fight this degradation, reverse entropy locally (i.e., build complex structures at the expense of energy), and even detect and repair genetic damage. None of these things happen in non-biological nature.

Top-Down Biology & Technology

Life's ability to maintain and repair genetic information across the generations is clearly evident. But biological systems exhibit far subtler mechanisms for managing information as well. This seems to be a distinguishing characteristic of life. "Despite the notorious difficulty in identifying precisely what it is that makes life seem so unique and remarkable, there is a growing consensus that its informational aspect is one key property, and perhaps the key property," writes Sara Imari Walker. "If life is more than just complex chemistry, its unique informational aspects may therefore be the crucial indicator of this distinction." The conventional scientific approach is all about reductionism: We might start by considering activity at the simplest or lowest levels, such as the interactions of molecules, and then we'd follow those effects "upward," to gain insight on how more complex (or higher-level) processes are driven by the influence of the lower systems. That is a bottom-up approach; it's about breaking down complex systems into their smallest elements, and then figuring out how those building blocks come together to produce the complexity. Ultimately we would like to understand how the highest-level process of all — consciousness — emerges from these low-level interactions. Recall that similarly, in the bottom-up approach to cosmology, we try to understand how the initial conditions of the universe lead to the present "fine-tuned" conditions, or we try to find some master law that causes a beginning universe to turn out like our present universe. But, like Stephen Hawking's top-down approach to cosmology discussed in the previous chapter, in recent years there has been interest in the top-down management of information in biological systems.

The top-down approach to biology acknowledges that other influences are at play beyond the brute interactions of particles. For example, complex processes can influence simpler, lower-level processes. A familiar example: When you think about something that makes you nervous, your physiology changes. Your heart rate increases as your nervous system tells the muscle fibers of your heart to contract more frequently, a mechanism that occurs on the molecular level. A conscious thought can alter molecular-level activity — that's top-down information flow. Biological systems are rife with incredibly complex flows of

Maxwell's Demon

What would it take to violate the second law of thermodynamics, which states that entropy in the world always goes up? The physicist James Clerk Maxwell imagined a tiny mechanical "demon" that could automatically select which gas molecules to let pass through a trap door. If the demon let only faster-moving molecules through, the gas on one side of the door would get hotter and hotter, with the other side getting colder and colder, thus reducing entropy. Would it be possible in principle to build a Maxwell's demon — something that reduced entropy in such a manner "for free," thereby violating the second law?

Theorists have been discussing Maxwell's demon for decades, and the consensus is in: As with life, the demon could only achieve local entropy reduction by increasing the entropy overall in the world. For example, in order to judge the velocity of different molecules against each other, the demon would need a memory function of some kind. Gathering and storing this information, and erasing the memory to make way for information about new molecules, would require the expenditure of energy, and therefore would increase the entropy of the world. And that entropy increase would *always* equal or outpace any entropy decrease the demon could achieve. Maxwell's demon is effectively a perpetual-motion machine, and as with those, you can't get something for nothing.

A connection with what we've discussed thus far: For a Maxwell's demon (or anything for that matter) to have a memory function, it must compare values measured at different times. This means creating durable relations across time (Chapter 7) that are truly informational: They assist in reducing the demon's uncertainty regarding the molecules' relative velocities. What systems are capable of creating, storing, and comparing informational relations across time? Biological systems, and the machines they build, seem to be the only contenders in the universe. In no other systems are relations across time durably stored, compared, and put to work as some causal influence.

information, which can operate in both top-down and bottom-up directions. "It is the interaction of bottom-up and top-down effects that enables the emergence of true complexity," writes the cosmologist George Ellis in a paper on this topic. Ellis points out that top-down causation does occur in non-living physical systems, for example the synthesis of heavy atoms in a supernova depending upon macro-level variables such as the density of matter. But complex *bidirectional* causal networks — involving the flow of information in both bottom-up and top-down directions — are exclusive to biological systems, writes Ellis. As with genetic sequences, these networks are maintained by homeostatic and repair mechanisms, a demonstration of how biological information endures in the world in ways that non-biological information does not. And ultimately, the genetic code is what makes all of this happen. Biological information includes information on how to ensure that the information endures as information!

"The fact that all phenomena of life are based on information and communication is indeed the principal characteristic of living matter," writes the physicist and biologist Bernd-Olaf Küppers. "Without the perpetual exchange of information at all levels of organization, no functional order in the living organism could be sustained. The processes of life would implode into a jumble of chaos if they were not perpetually stabilized by information and communication." Sara Walker and Paul Davies point out that in biological systems, the various chemical reactions are "orthogonal," meaning they stay out of each other's way and don't mutually interfere, which would be disastrous even for a cell's most basic functions. Walker and Davies mention that analog computers were replaced with digital computers, in part, because it is easier to keep various channels of digital information out of the way of each other, whereas analog circuits are subject to contaminating effects such as induction, cross-talk, and other interference. "Orthogonality is, by comparison, relatively easy to achieve with digitized switches," they write.

However, no one knows how such orthogonal digital switching and genetic coding could even get started in a prebiotic world of stuff — especially when configurations of nonliving matter tend to degrade due to the unrelenting effects of entropy.

With these kinds of analyses, it gets convoluted to assume that biological organisms are fundamentally made of stuff, where the long-term integrity of their stuff-systems depends on flows of information that derive from the stuff itself. It is a less complicated picture to imagine the informational flows as the fundamental processes — however complex these flows might be — with the intricately interacting stuff emerging, secondarily, when we conscious, intelligent humans stick our noses into nature and interpret what we see.

It's worth pointing out that pretty much *all* biological information is digital in some respect. Consider chemical signaling, for example by a hormone. When we look at how a hormone acts on a cell, we see that the substance both goes out and is received in the form of quantized packages that we call molecules. And, half a molecule won't do the job. This quantization of cause-and-effect is how physiological interactions happen in general: There are few or no cases where one part of an organism signals another part on a continuous, sliding basis that can take any value, the way the angle of a board with a ball rolling on it can take any value. For example, nerve impulses passing along neurons are described as spikes of electrochemical potential that occur in the form of discrete packages. Nerve cells do not convey information by way of a smooth, continuous change, like a hose under air pressure, or a light bulb connected to a dimmer switch. There are no signals or switches, anywhere in organisms, that act like smoothly turning analog volume knobs. Information flow in biological systems seems to be discrete through and through, just like the discrete information of digital bits.

There's one other type of complex system I've skipped over: technology. Indeed, there are many technological systems that register, process, and durably store information. I'm writing this book on one such system. But — and here comes another key concept in the simplest-case scenario — technological systems ultimately *are* biological systems. We biological beings design them and we build them; there would be absolutely no technology on Earth were it not for biological organisms. Technology demonstrates top-down causation: Consider a computer that measures the number of seconds since the most recent keyboard or mouse activity, compares that number to a preference parameter, and when the numbers are equal, directs electrical signals to the monitor to display a screen saver. The screen-saver program, in turn, was designed by conscious humans writing lines of code, another top-down action. Technology is an extension of biological capabilities, and many kinds of technology register and store information about the world, as seemingly received and delivered by photons, molecules, or other particles "out there," and their related forces. We designed technological scientific instruments to be extensions of our senses, so that we could better measure our world.

Even a photographic plate in a photon experiment is a human-prepared measuring tool, which we place in the path of a beam of light that we also prepared, for the purposes of detecting photons. Having queried nature in this manner, we shouldn't be surprised to find the appearance of information — in this case, specks indicating the registrations of photons — after we dump the plate in developing fluid. We should not be surprised that our tool for registering photons has accomplished the task it was designed to do: to acquire information about photons, in the same way the eye does, after we had set up an experiment to prepare light and detect its particles. For these reasons, biological organisms and their technological offspring can be combined into one class: *techno-biological* systems. The manner in which all techno-biological systems register, store, and share information, as durable relations across both space and time, is essentially the same — while also being completely different from anything that happens in non-living nature.

If information is fundamental to the universe, and if techno-biological observers are uniquely capable of registering, storing, and using information, then we have a way to explain not only John Wheeler's it-from-bit conjecture, but also Robert Lanza's idea that life creates the universe, not the other way around. The information we have gathered about the world is relative *specifically* to us, and by extension, our technological devices. Since we techno-biological observers are ultimately made up of nothing *but* information, it would be accurate to state affairs this way: The universe consists *only* of the informational relations that make up techno-biological systems. There is nothing else in the universe. That is why it is so much simpler than we think.

The Problem of Multiple Observers

This brings us, finally, to the last piece of the puzzle for the simplest-case scenario. There is one feature of the world, more than anything else, that makes the "stuff" approach seem deeply intuitive. It is probably the chief factor that got Western natural philosophy on the wrong track, going back to the ancients, and so it becomes a problem for anyone arguing that the world is not a collection of stuff out there in space.

The problem is this: Why do multiple observers see the same world, with the same course of events? If a rock is *not* made fundamentally of stuff out there in an objective physical space, why do you and I observe the same rock, with all of the same properties? The solution turns the simplest-case scenario from a curious but incomplete conjecture about the nature of things, to an all-encompassing vision, one that shows how the world works from the finest levels to the grandest.

Stepping back for a moment: I've argued that the informational approach presents a simpler picture of the universe, because to produce the same observable world, it requires a minimum number of individual fundamental entities (perhaps 10^{50} bits, a guess) that's smaller than the minimum number required in a vast universe of defined stuff (perhaps 10^{80} particles of different kinds, or more). But there's also the issue of the number of observers: The stuff approach assumes that there are many observers, each fundamentally consisting of an independent bundle of stuff, separated from other observers in space and time: seven billion humans, far more insects, and something like 5×10^{30} bacteria, all gathering information about the world in their own way. But what if there were an approach where these different observers actually constitute only *one* big observer?

The Topology of Life

Topology is a branch of mathematics that deals with the fundamental similarities and differences between shapes or structures. To many people, topology is the discipline where a coffee cup is the same as a donut. The idea is that an ideally pliable donut could be stretched and molded into the shape of a coffee cup, without any cutting, breaking, or joining, since the donut's hole can be repurposed into the hole in the cup handle. For a hole-less croissant, however, this wouldn't be possible, as you'd have to break into the pastry in order to make a handle. Thus, topologically, a croissant is fundamentally different from a donut or a coffee cup. Topology also deals with *connectedness* — one donut is fundamentally different from two donuts — and *compactness*, where a donut is described as compact because it is a closed shape that includes its boundary or edge. Topology has many applications, from electronics to cosmology.

Think about the long line of ancestry that I discussed at the beginning of Chapter 6. From every generation in that lineage to every next generation, there is a direct chain of causation, an informational continuity from one generation to the next, genetic information being only a part of this connection. A mother physically gives birth; I cannot think of a more extreme case of causation than that. Particularly also in the case of mammals, there are nurture influences as the offspring is raised, by parents and perhaps grandparents and/or other members of an extended family, in the case of highly social mammals or social insects.

This chain of information and causation is unbroken, extending across the generations from yourself to any ancestor you might arbitrarily choose. Indeed, this is true of any living organism. In no case does an organism appear spontaneously without an informational, causal connectedness to at least one parent;[2] Louis Pasteur put an end to that myth of spontaneous generation. And the incredible thing is, if you step back and imagine *all* of the lineages of *all* of the biological organisms that have ever lived, you realize that this unbroken connectedness extends across the entire history of life — just like that one spindly thread that extends across billions of years leading to you, today. The entire history of life can be treated as a single connected object.

Here's a thought experiment: Consider the causal connection between you and an ant. If you go back far enough in your chain of ancestry, you'll encounter an ancestor common to both you and the ant. In our modern description, it may have been something like a bilaterally symmetrical sponge-like thing 500 million years ago, give or take a few dozen million. Perhaps it had one offspring that was quite a bit different, and this one ultimately gave rise to vertebrates, while one of the others gave rise to arthropods. It's incredible, but this is how evolution works. So there's a common ancestor, and the chain of ancestry from that ancestor to you is continuous and unbroken, in a causal sense. Meanwhile, the chain of ancestry from the common ancestor to the ant is continuous and unbroken as well. That means the two causal chains are, in turn, connected to each other: The chain between you and the ant, back down hundreds of millions of years and back up again, is unbroken in its entirety. It is a single and continuous shape, although it is forked at the bottom (you can think of the "handle" of this fork going down further generations if you like). Topologically, the history of the two organisms and their

---

2 The only exception is the first organism, our earliest common ancestor. This appears to be unavoidable — there must be *some* causal boundary at the beginning of a non-eternal universe, whatever approach to the nature of things that you prefer. I refer to this as the logic singularity (see page 160).

You   **Ant**

ancestors is a single connected structure. The same principle of connectedness applies to any two organisms you might select, including yourself and a dinosaur, or a ficus and Albert Einstein. The continuous, unbroken structure of causality and information flow ends up including every biological organism that has ever lived, along with every technological device ever created. They, and we, are all connected in a profound manner, as an extremely intricate tree comprising the entire history of life (see **Fig. 2**).

What about non-living things? Can an argument be made that a human has an unbroken causal connection to a rock? Perhaps, under the stuff view of things, or at least under a "book-world" treatment of information, that follows the causal connections among all matter back to the Big Bang. However, the amount of information describing the history of the universe since the Big Bang — and describing every emergent particle in the observable universe, and beyond, that was in contact at the Big Bang — would be much greater than the amount of information that has been registered by techno-biological systems. If we can find a way to have the same observable universe with a much smaller number of bits, Occam's razor suggests that the fewer-bits picture is more likely the correct one. Also, the

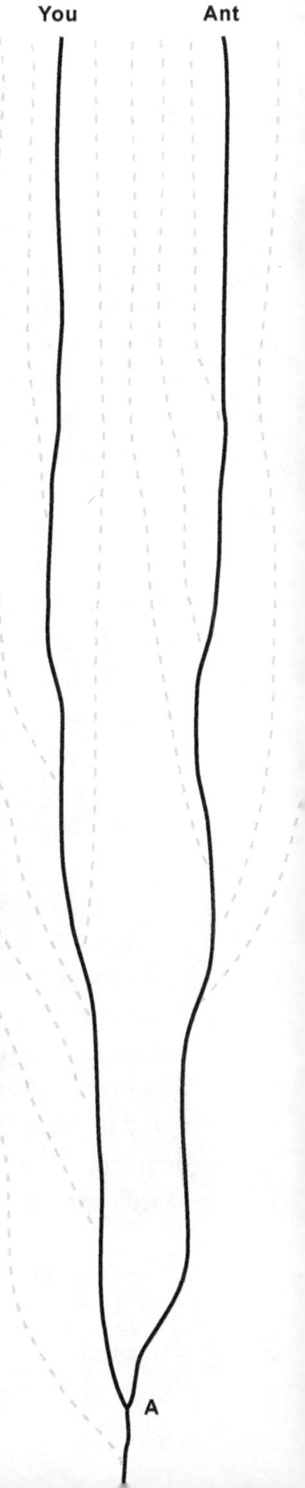

Fig. 2. The chain of causality between you and an ant (top) is a single connected informational structure, linked at the most recent common ancestor, perhaps 500 million years ago (**A**, bottom). The structure also includes every other lineage of every other organism that has ever lived on Earth (dotted gray lines).

A

tree of biological information and causality is truly continuous, the genetic information passing from parents to offspring being the clearest example. It would be harder to argue that the tree of causation for an entire universe of stuff is connected throughout, especially when random stuff-events, such as the spontaneous decay of atomic nuclei, are routine and seemingly built into the laws of physics. Finally, the system of techno-biological information is finite, with an open boundary at the present time, as information keeps getting added to the system. In contrast, a tree of causation for an entire universe of stuff could go on forever, and at least (presumably) extends beyond the observable universe, with no discernible boundary at all. This is a minor point that's admittedly aesthetic, but a finite, tightly connected universe is more elegant and beautiful than a possibly infinite universe of stuff far-flung across billions of light-years of space. Regardless, it is clearly a simpler picture.

Biological life plus its technology is unlike any other system or chain of events that we know of — not just in terms of complexity but also its unbroken connectedness, in causal, genetic, and gestational/nurturing/creation terms. But what is the significance of this connectedness?

Connectedness Means Constraint

Let's imagine some kind of object — a one-celled living organism, perhaps. If that thing were to split into two other things, then those two things will be linked: The three objects have a causal connectedness across the history of their existence, just as two organisms with a common ancestor do. Also, post-split, the two daughter-things share a mutual relation in space.[3] The two daughters also have relations across time with their parent-thing, as well as relations across space with the parent's original location.

---

3 This visualization is complicated by the need to establish some kind of reference frame in order for us to describe the things' relations with any meaning, as explained in Chapter 3. For simplicity, here we will assume a marked-off Cartesian grid and a ticking clock, but the principle I am describing applies in any reference frame.

These are all relations that we, as third-party observers, could measure and quantify relative to standard measurements that we've devised, as discussed in Chapter 3. And, in performing these measurements, we would notice that the relations always remain consistent with one another as parts of a larger system, not unlike the system of relations of debts among the four friends on page 105: If one element changes, the relations all change accordingly, with internal mathematical consistency. This constraint can be shown geometrically:

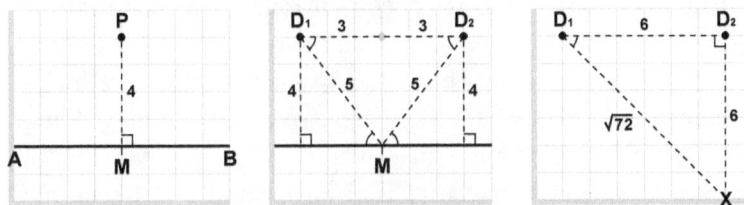

Fig. 3. Consider a parent-thing (**P**, left) measuring 4 units to a perpendicular line segment **AB**, at its midpoint, **M**. If **P** splits into daughter-things **D₁** and **D₂**, any subsequent measurements will be mathematically consistent with the previous relation (center). If one of them then got new information on an apparent object **X** (right), the other daughter's measurements would be constrained such that **X** appears to be in the same place. This is due to the **D₁-D₂** system's legacy information on their spatial relation with each other.

Constraint is something that happens when two objects are linked in a causally connected manner. Whatever one of the daughters might observe, is logically required to be consistent with whatever the other daughter might observe, given their mutual history such as relative motion with respect to each other. Throughout this book I've talked about old or legacy information, and how such information imposes constraints on a system. This geometrical example demonstrates the effect nicely: Since the daughters share a history that connects back to the parent, each daughter's measurements must constrain the measurements of the other. Logical consistency — what might be called the observation of a real world according to the definition at the end of Chapter 1 — is enforced within the structure that has this property of causal connectedness. The structure of the parent-thing plus the two daughter-things constitutes a world of logically consistent observership. So, three observers that seem to be disconnected in space and time are actually required to observe the same logically consistent (real) world, even in the case that such a world and its observers aren't made of stuff "out there."

What Mechanism Enforces Logical Consistency?

It's easy to understand why an object would be viewed the same by two observers, if the object (being made of material stuff) is fundamental to the world. It's not so easy to understand, though, why two observers might be constrained to receive the same information, in a world that is fundamentally informational. How is this constraint enforced? What force is responsible for making sure that the world remains logically consistent?

In the case of causally connected observers, it appears that things just could not be any other way. If you measure a particle as spin-up, logical consistency necessitates that another techno-biological observer will also measure that particle to be spin-up. Why? For the same reason that in **Fig. 3**, if D_2 measures a distance of five units to the midpoint M, D_1 will also measure five units. A longer or shorter measurement would be a geometrical contradiction. Also for the same reason, if you hold up a coin and someone on the other side of the coin sees heads, you will not also see heads (unless a mirror is involved). As another example, a half-man, half-antelope "mantelope" could never be produced by two antelopes (at least without extraordinary genetic engineering). Logical consistency in the world is a simple fact of nature, and it has never been observed to be violated. So if there were *any* assumption we could make when speculating about the ultimate nature of reality, logical consistency would be a good candidate. Nature abhors a contradiction.

Connected Constraint Means Unity

For any causally connected system, different observers that are part of the system — or perhaps better said, different observing subsystems — will observe the same world. That world is real in the sense of Chapter 1, where the internal system of relations between elements is invariant: The relations, as well as the relations between the relations, would not change even if you stretched them in some arbitrary but uniform manner (see page 27). More important, anything that happens to one observer in the system constrains what happens to all observers in the system. If one daughter-thing in **Fig. 3** moves to the left, that will change its relation to the other daughter, and therefore the relations resulting from all other measurements in that world will change accordingly. What this internal consistency means is, a system of observers connected in this way can be treated as *one* single observer. It is a kind of *super-observer* that comprises many individual observing subsystems, which are tied together in mutual informational constraint. And that, in one sentence, is the simplest-case scenario for the universe: The universe is a finite system of information relative to one super-observer, that super-observer comprising all information registered, either biologically or technologically.

One beautiful thing about this picture is that it is fractal (see page 136). The techno-biological super-observer can be understood at multiple scales, including the intuitive human scale. Think about your physical body: It has numerous senses to pick up information in different ways, and they're in different locations in space. If you pick up and eat a piece of cantaloupe, your hands are feeling it, your eyes are seeing it, your nose is smelling it, and your tongue is tasting it. Your ears can even hear the texture by your chewing. Yet, you have no problem seeing yourself as a single observer of this cantaloupe. Of course, the fact that all of these observing sense organs are within the same skin helps to create that impression, as does your continuous identity as an individual person, and the fact that the sensations are coordinated within your brain to produce a whole that constitutes the conscious experience of you eating a cantaloupe. But if we look at the information flows, we see something like a tree structure, with the sensory information coming together from numerous sources and interacting in the brain. Multiple observing systems in your body are measuring the cantaloupe in different ways, but in the larger scheme of things it's really just one observer — you.

Consider also information flows on the cellular level. A cell may have different ways of registering information from the environment or from neighboring cells, and different subsystems for responding to this information, which in the extreme case includes cellular division (a process not unlike the splitting of the parent in **Fig. 3**). The cell contains numerous observing subsystems, some of them embedded in the cell membrane the way our touch receptors are embedded in our skin. However, the cell can be considered to be a coordinated observer in its own right, one of many that make up an organism. This analysis may be continued down to the molecular level, for example, in the subsystems that coordinate in the transcription of RNA and the synthesis of a protein. This machinery can be seen as part of a larger subsystem (e.g., a cell), which performs tasks as part of a still-larger subsystem (e.g., an organism).

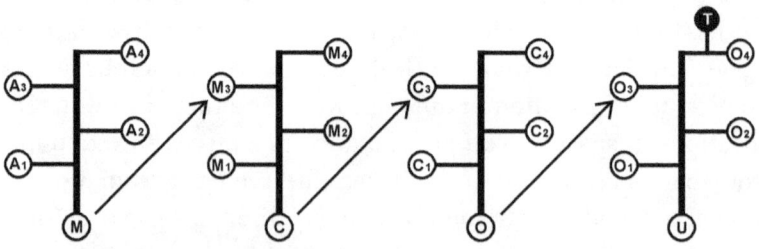

Fig. 4. The fractal nature of informational subsystems: Molecular processes (A_1–A_4) make up a cellular mechanism, **M**. Such mechanisms, in turn, make up a cell, **C**. Cells make up an organism, **O**. All of this plus technology, **T**, make up the universe, **U**. In each case, the component subsystems are part of a larger subsystem, which may be nested in a still-larger subsystem.

It's not always clear what the word "organism" means. For example, it is quite natural to think of an individual honeybee as an organism, since it's analogous to other organisms, being bounded by an exoskeleton, and so on. However, one can easily imagine a colony of honeybees as the actual organism, given how each bee has relatively simple functions, and hundreds of them cooperate to produce emergent behavior that is more complex than that seen in any of the individuals. When a colony of honeybees gets too large, it may split into two colonies, one of which swarms and goes looking for a new home. Individual scout bees explore locations and bring information back to the swarm, which makes a decision by consensus of where to set up the new hive. In describing this behavior, it's difficult not to use singular-verb phrases such as "it goes looking for a new home" — as if the swarm is a single thing. Such collections of animals are sometimes described as *superorganisms*. As a different example, it may seem peculiar to us that a large stand of aspen trees can really be many sprouts of one connected plant, all of them genetically identical. But the fractal nature of the simplest-case scenario makes all of these kinds of distinctions somewhat arbitrary, just as it can be arbitrary to look at one of those aspen trees and decide what exactly constitutes a branch.

Every true fractal has an ultimate structure, and if we step back and look at the big picture, we see the biggest superorganism of them all, which embraces every informational substructure in the world: All biological organisms that ever lived, plus all of the technology we have ever made. This ultimate super-observing superorganism is ruled by constraint: Any information registered by an ancestor, or by a biological cousin, no matter how distantly related, constrains any information that you or I may ever receive. If an ancient warrior carved his name on a wall, or a tree root split open a face of granite, then we too will find that carved stone or split granite when we go looking for it. But not necessarily because the world is made of stuff. Strictly speaking, even a stone carving may not be carved in stone!

Living Matter Flows Like a River

One of the more peculiar aspects of biological systems is that there's a distinct continuity of information, even as atoms and molecules come and go. Our bodies are constantly recycling themselves, breaking things down and rebuilding. If you could compare the stuff making up your body now to the stuff from several years ago, you'd discover that very little of "you" is left. And yet, you're the same person, with the same memories and other aspects of your identity that you had all along. Each of us is like a river, whose course remains essentially the same, even though the water passing through changes incessantly.

The continuity of identity is a discussion topic among philosophers of the mind. Identities happen in animals, which can maintain specific memories for years. If "stuff" is fundamental, and the information that makes up one's identity derives from stuff, then our identities must be reducible to the configuration of matter in our brains, alone — ultimately there must be no other contributing factors. But this would be like saying that the water in a river *alone* is responsible for the course that the river takes, when in fact the waterway is shaped by the more permanent spatial relations between the river's banks. The river's "identity" results from the stable relations that channel and direct the water; it does not result from the water-stuff alone, which is just going along for the ride. Similarly, the information from our genome and our environment is what ultimately determines where stuff in our body goes and how it is manipulated. These things aren't up to the stuff itself.

In the simplest-case scenario, matter is emergent, and in biological bodies, the factor that channels this emergent matter is information, for example as seen in the synthesis of proteins from RNA. In biological systems, the emergent stuff may come and go, but the informational relations — including those across time — remain durable. This is another indication that the complexity of life is more than just the complex sticking-together of atoms.

Informational constraint applies to technology as well, all technology having been created and built by tool-making humans. Technology that's designed to measure — particle detectors, electronic scales, seismographs, the list goes on and on — is all fashioned after our own sensory functions, which query nature and make measurements by comparing observations to known relations (as discussed in Chapter 3). Being part of the super-observing superorganism, a technological device, upon making an observation, will constrain the observations of all other connected observers, whether biological or technological. And, just as we can arbitrarily decompose any organism into subsystems, such as sensory organs, or even individual cells of sensory organs, similarly we can decompose technological devices into their constituent subsystems, and treat those subsystems as observers. A desktop calculator is human-designed to observe button-presses — the mechanism under each button can be thought of as a separate observing subsystem — and it also has memory functions, which allow it to create durable relations across time, observe the values in its memory functions, compare them, and respond appropriately. Even though technology is vastly less complex than biology, the observing and measuring functions are remarkably similar.

When I claim that the universe is ultimately made of biological organisms plus our technology, it may sound like I'm saying there's something special about, say, the matter in the body of a turtle versus the matter in a rock. No! We are taking an informational approach to the nature of things. And in this approach, as John Wheeler conjectured, *all* matter emerges from information. That means the matter that makes up biological organisms, the matter that makes up technological devices, and the matter that makes up everything else, all emerges in the same way, from information. The simplest-case scenario claims that this information, and I mean all of it, is exclusive to the super-observing superorganism. The matter that's observed to make up rocks and water and distant quasars, and indeed the observed matter of our own physical bodies, is all the same kind of matter. But it all emerges relative to one, single system, and that system is the techno-biological superorganism, that extremely complex tree of informational relations which has evolved over billions of years. The world *is* this system of techno-biological informational relations, and nothing else; all of the matter and energy, every physical law and every field, emerges relative to observers that are part of the superorganism. This is how the universe works!

The superorganism discovers the universe kind of like the way an octopus might discover a Christmas present. One arm might shake the box, another might weigh the box, maybe another has an x-ray machine or CAT scan. Another arm might peel off a corner of the wrapping. Slowly, information about this present appears to the octopus. And, as each piece of information comes in and adds to the whole system, the octopus' uncertainty about the present goes down. Fully unwrapping the present, and learning everything possible there is to know about it, would be analogous to Paul Davies' notion of filling the universe with observations. We haven't done that and we will never do that, so as a super-observer, we continue to gradually "unwrap" the universe, bit by bit, over the course of billions of years. (The obvious weakness in this analogy is that the universe's information doesn't pre-exist in the way a wrapped present pre-exists.) Our uncertainty about what the world might be like diminishes as the informational content of the world grows with each observation.

This is how the universe evolves *in parallel* with biological life evolving. The super-observing superorganism complexifies, and equivalently, so does the emergent universe observed by the superorganism. Viewing the universe as something that evolves in the same way that life evolves, we find parallels between life and the informational universe:

1. The history of life is like a tree, sometimes called the tree of life, with simplicity at the bottom and complexity at the top. The informational universe also has a similar tree-like structure.

As informational relations accumulate on the tree, the context-enhanced meaning of newly appearing information goes up, relative to subsystems[4] on the tree sufficiently complex to process this information. This is how complexity and meaning in the universe develop, as discussed in Chapter 5.

2. In the evolution of life, humans are highly complex organisms that have developed the capability to communicate through language. This mirrors the general evolution of the universe, which has developed complex subsystems with the capability to register, process, and share large amounts of information with other subsystems. These subsystems can be human neurons, regions of the human brain, humans, groups on Facebook, parts of a computer, computers, networks of computers, etc. The universal applicability of the same principles across various scales and realms is partly what makes the simplest-case scenario so simple.

3. As discussed at the beginning of Chapter 6, the course of the world appears to have been fine-tuned to produce you, since an extraordinary chain of events over billions of years had to occur just so, in order for your ancestors to survive and for you to be here today. But this is only a consequence of interpreting your existence today in the context of everything that you know could have gone wrong in this long process, but didn't. Similarly, the universe appears to be fine-tuned to produce matter and life. But this is only a consequence of interpreting the existence of matter and life today in the context of all of the physical parameters that we now know could have been different, but weren't.

4. In natural selection, organisms that are less fit are removed from the gene pool. Over the course of millions of years, the lineages that survive become increasingly fit, to the point where today, those lineages that remain appear to be amazingly fine-tuned for their environment. Similarly, as the universe evolves and

---

4 A subsystem can be an organism, a colony of organisms, an organ of an organism, a single sensory cell, a computer chip, etc. — it's any part of the superorganism capable of registering and processing information.

That Cat of Schrödinger's

The thought experiment known as Schrödinger's cat is a famous challenge to quantum mechanics: A cat is in a box, along with a radioactive sample, particle detector, and flask of poison. If an atomic nucleus decays, the cat is killed. Without our observing whether or not one or more nuclei have decayed (by opening the box), quantum mechanics describes each nucleus as being in a state of *superposition*, both decayed and not decayed. Presumably, though, that also means the cat is both dead and not dead. Erwin Schrödinger meant the situation to sound absurd, and this paradox-of-sorts is resolvable in a number of ways, depending on the QM interpretation applied.

In Carlo Rovelli's relational quantum mechanics (pages 120–129), the state of a system is literally the relation between that system and whatever observer we choose. The particle-detector + cat system can be an observer, so relative to it, the nucleus may have decayed — even though to the person outside the box, it remains in superposition (which in RQM really just refers to informational isolation and the resulting uncertainty).

The simplest-case scenario takes RQM one step further. It treats the cat, the particle detector, and the person, or any combination of them, as subsystems of the techno-biological super-observer. Therefore, even if the person doesn't know what happened inside the box, his or her subsequent observations (or those of any other living thing or device) will necessarily be constrained by any observations made by the detector and/or cat. The same would be true if the cat were not there — or were replaced by a balloon at the mercy of a pin — since the particle detector alone, being part of the same techno-biological super-observer, would be equally effective at constraining events affecting living and technological observers alike, everywhere.

In principle, experiments could test whether or not such constraints can be set up between techno-biological observers, and so could test the validity of the simplest-case scenario (see the Appendix).

techno-biological subsystems make observations, potential histories that are in contradiction with these observations get culled away. After many millions of years of this culling, the remaining potential histories strongly contrain observations, which creates the impression that the universe has been fine-tuned for matter and life all along.

I recognize that these points have a somewhat circular, self-referential quality: Life evolves the way the universe evolves, because the evolution of the universe is the evolution of life. However, this may be just the consequence of an informational universe that creates itself — pulling itself up by its own bootstraps as it were, with no help from any outside agency. To again borrow Paul Davies' words, such a looping feedback-like process is necessary for "a universe that has an explanation within itself … because there's no other way you can have an explanation for the universe from entirely within it." The evolution of life and the sharpening of the universe aren't merely similar. In the simplest-case scenario, they are one and the same process.

Entanglement Explained

If you'd like some experiments to demonstrate the constraint that happens when systems are causally connected — I wouldn't blame you — look no further than the experiments involving quantum entanglement, such as those discussed in Chapter 2. (The most famous one was performed by Alain Aspect in 1981.) When two particles are produced together in a single event, their intrinsic spins will be found to be oriented in opposite directions, or anti-correlated, when the spin for both particles is measured along the same axis. To be consistent with the laws of physics, these spins have to cancel each other in accordance with the conservation of angular momentum (see pages 52–53). The difficulty with entanglement has always been how two particles, even if they're a great distance apart, can maintain their anti-correlation, as measured along any arbitrary axis. They seem to be signaling one another — faster than light, no less: "I'm being measured along *this* axis, and I'm oriented this way, so you need to be oriented *that* way." This is another example of our intuition misleading us, since pretty much no one believes that this kind of communication between particles takes place.

Now that we understand causal connectedness, and the way that connectedness constrains what can happen to all members of the connected system, the phenomenon of entanglement makes perfect sense. In fact, it couldn't be any other way! The entangled particles emerge from information that derives from a common causal origin. This connectedness constrains what can be discovered about one emergent particle relative to its entangled twin. If we get spin-down information associated with one of these emergent particles, the other must be spin-up; if one is measured to be spin-left, the other is sure to be spin-right. There is no better experimental demonstration of the constraint that I've written about in this chapter.

Let's look at this from another angle. Generally in science, we reflect on the laws of the universe that we know, and we use that knowledge to set up experiments to observe certain things, to collect information about the world or perhaps to probe the physical laws more precisely. We perform measurements, and new information appears, in accordance with those laws. We then interpret the results — even very simple, one-bit yes–no, up–down values, like the one described at the beginning of Chapter 1 — through a rich contextual lens. There are the laws of physics and the Standard Model, and there's newer information about the way we set up the experiment, and finally, we have new information from the experiment's results. Combining all of this in our consciousness or notebook results in a rich narrative of particles buzzing through space and even seeming to signal each other. But really, the new information that's appearing in the world, such as the result of a spin measurement, is simply constrained by the information that the techno-biological superorganism already has. This includes the results of spin measurements performed on entangled particles.

And those informational findings will *always* be consistent from observer to observer — every connected subsystem will see the exact same world of emergent stuff, guaranteed.

Entanglement is particularly amazing, because it demonstrates several principles of the simplest-case scenario:

1. Information about spin appears in the world when it is sought out by techno-biological observers. The information comes about wiki-world-style, rather than being a pre-existing, persistent feature of the world carried by particles, book-world-style.

2. The space separating the particles is not a fundamental feature of the world. Spatial separation emerges out of the information that we observers have (from which also emerge the particles themselves) — as interpreted in the context of all prior observations (from which emerge the laws of physics).

3. Partial information has the effect of partially constraining subsequent measurements; complete information completely constrains those measurements. A definite spin-up measurement of one particle means that its entangled twin is guaranteed to be found to be spin-down. But if we measure the twins along slightly different axes, the second measurement will be only partially constrained, by a degree that is predictably statistical (see pages 60 and 94). The experiment's probability-driven nature is counterintuitive if we assume that the world is made fundamentally of stuff, but it makes sense in the constrained-information wiki-world picture.

4. Like biological observers, technological observers such as Stern–Gerlach apparatuses are subsystems causally connected to the super-observing superorganism. We can use a different apparatus for each measurement and the result will be the same. This is a good example of different observers always having descriptions of a system that are consistent with each other, even though those descriptions may not necessarily be identical (Carlo Rovelli's insight, pages 120–129).

An entanglement experiment is somewhat like two distant observers looking at the same coin through telescopes

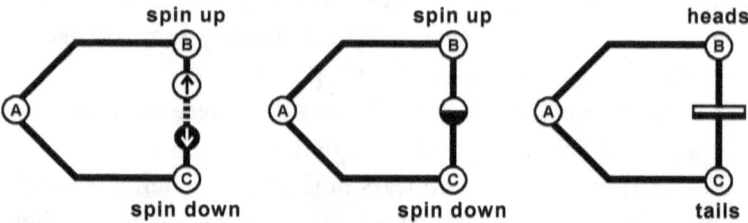

Fig. 5. For two observers **B** and **C**, connected by a common ancestor **A**, the spins of entangled particles are anti-correlated (left). But these relations stem from a single informational source (center), as if the observers had looked at the opposite sides of the same coin through telescopes (right). In one case, the information seems to be carried by the emergent particles themselves; in the case of the coin, emergent photons seem to carry the information.

(see **Fig. 5**): If the coin is facing flat, plain logic dictates that one observer will see heads and one will see tails. You can appreciate that both observers seeing heads would be a contradiction — such a thing could not happen in a logically consistent world. The observers will always get anti-correlated measurements, out of what is essentially a single bit of information (the either–or direction that the coin is pointing).

If you think through the delayed-choice and quantum-eraser experiments from Chapter 2, you can similarly understand that the peculiar findings — which are counterintuitive and paradoxical in the book-world view of stuff — are in alignment with the picture of a super-observing superorganism gradually receiving information about an internally consistent world. New information appears when some subsystem of the superorganism asks nature a question, and the answer is never in contradiction with the information the superorganism already has. When we talk about the world or the universe, we're really talking about all of the informational relations that the super-organism has built up over the eons. Since the superorganism itself ultimately consists of nothing *but* this internally consistent set of relations, the universe, quite literally, *is* the super-observing superorganism.

I shudder to put it this way, because it sounds like it's straight out of pseudoscience literature, but an accurate way to describe the connected constraint I've discussed is: We are all, all of us, quantum-entangled. When something happens to one of us, it happens to all of us, even though we may not know it. To express the situation in the most poetic language possible: *When you gaze upon a rose, so doth your distant love.* (Then again, so doth every frog and every mushroom and every Roomba, in their own way, but that's a little less romantic.[5])

Information Gain vs. Constraint

There is one last technical matter that needs to be addressed. (After this section, you won't have to think about experiments or entanglement or anything like that.) An observer being constrained, by the presence of information in the world, is not necessarily the same as an observer *receiving information.* The distinction is nuanced, so bear with me as I tease it apart.

If I measure the spin of a particle along the up–down axis and find it to be spin-up, that is the receipt of information. It's a reduction of my uncertainty regarding this question, which I posed to nature by configuring the experiment just so, orienting the apparatus in the up–down direction and so forth. However, when I receive spin-up information, I don't know whether that measurement result was a random 50/50 outcome, or whether the result was constrained to be *necessarily* spin-up. Recall that measuring spin of a previously unmeasured atom is a matter of probability, like a coin-flip. But if a colleague had already measured the atom and found it to be spin-up along the same axis, or if a friend had measured the atom's entangled twin and found it to be spin-down, then the atom I'm measuring couldn't produce any other result. It *has* to be measured as spin-up; that information already exists in the world, as measured by another connected subsystem, so I am constrained to receive

5 As is the notion that you and your love are, ultimately, genetically related.

the same information — or the opposite, in the case of two en-tangled particles. Yet, *if I do not know* about previous related measurements, I cannot tell the difference. By a single measure-ment in isolation, I cannot determine if my result is random or constrained, any more than I can determine whether or not a coin or die is fair by flipping or rolling it only once. I can learn only so much with a single measurement.

Knowing whether an experimental result is random or constrained requires additional information. Some kind of com-parison needs to be made. For example, if I'm rolling a die, I can compare the first roll to the result of many other rolls and get a good statistical idea whether or not the probability of rolling any particular number is exactly one-sixth. In that case, I'm combining multiple pieces of information to create a collective context in which to interpret the fairness of the die. This is analogous to the experiment where many photons accumulate on a screen, and then we use this mutual context to indirectly measure the photons' wavelength (see page 100). In the case of a friend already having measured the particle or its entangled twin, in order to learn that my measurement was constrained rather than random, I need to compare my result with my friend's result — including the contextual information associated with this result, such as the axis of measurement. There are two observers here, and the observers have to come together in some sense and exchange information in order for any correlation or anti-correlation between the measurements to be revealed. Like the example of rolling a die many times, additional information beyond the first measurement is needed in order to detect the presence of constraint and therefore have information *about constraint,* above and beyond the mere information *about the spin* I measured. That's what knowledge is: It's enriched information that reveals broader, contextualized connections among things in the world.

Constraint is literally instant and immediate, across the entire universe. When one component of the techno-biological superorganism receives information, the entire system becomes constrained. The constraint does not need to propagate across space, like a wave. We know this is true from entanglement experiments, and we can also reason that if it weren't true, logical contradictions could appear in the world. Suppose I measure the spin of a particle to be spin-up, and then you, seven miles away, measure its entangled twin one nanosecond later.[6] If we had to wait for the constraint to propagate through space from me to you, you might measure the particle also to be spin-up in the meantime, and having two spin-up particles from the same source would violate the conservation of angular momentum. Surely enough, such contradictory results are never seen to occur, and this is empirical evidence that the constraint of entanglement does not race across space at a finite speed. A physicist would say that this effect is *nonlocal*: It exhibits "spooky action at a distance," as Einstein described the faster-than-light interaction that seems to happen between entangled particles. Naturally, in the informational view there is no spookiness or communication between stuff-particles.

In contrast, the passage of information from one observer to another *does* proceed through space at or below the speed of light. Two people may separately measure the up-or-down spin of two entangled particles, and those measurements will be anti-correlated. However, in order to verify the anti-correlation — to gain knowledge of the constraint, for their uncertainty regarding *that* question to be reduced — they need to compare notes. This comparison is something that must happen along a point-to-point information transfer, like a telephone call, which is limited by the speed of light. It is an ordinary, "local" interaction, like all kinds of communication we're familiar with. It would obviously be a useful technology if

---

6 As timed by a clock midway between you and me, with all three systems at rest with respect to each other. Special relativity requires the specification of this reference frame, because phrases like "one nanosecond later" or "simultaneously" become undefinable if the clocks and observers are allowed to move around relative to each other.

entanglement could be used to transmit messages faster than light, but this appears to be impossible, according to decades of research. Even though spatial distances create no impediment to constraint, and observers everywhere can become constrained without their knowledge, an observer's constraint *alone* cannot reduce their uncertainty regarding anything — and that is why entanglement can't be used to transmit information. I have proposed an experiment to demonstrate that spatially separated observers can be mutually and nonlocally constrained, without the involvement of entangled particles; see the Appendix.

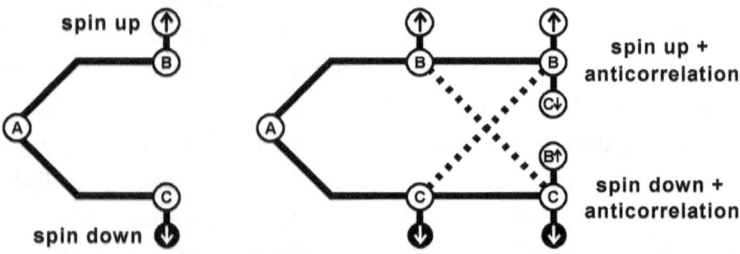

Fig. 6. Two observers **B** and **C**, connected by a common ancestor **A**, measure particles to be spin-up and spin-down, respectively (left). The spin of their own particles is the extent of their knowledge. To gain knowledge about anti-correlation, they must communicate or transfer information across space (dotted lines, right). Such information transfer is limited by the speed of light.

Going over this distinction again: When an observer makes an observation, information appears. However, it's impossible to know whether that result is truly new information in the world, or whether it's constrained by the action of some other observer. If the result was constrained, the only way that this can become *knowledge* is if the observer gets further information. Constraint by itself cannot reduce uncertainty; only information acquisition can do that, and the transfer of information from one point to another is limited by the speed of light. I suspect that this distinction has deep ties to an emergence of time and space in tandem, time–space equivalency being intimately tied to the speed of light — but that topic will have to wait for another day.

Our world is a wiki-world (Chapter 2) in which all of the "contributors" are anonymous. It is a cosmic melting pot of information, constraining all observer-participants, as John Wheeler called them, in the same manner. Meanwhile, observers are free to exchange messages and disseminate this information, leading to contextually enriched descriptions of the world for larger, more complex subsystems capable of taking part in this informational dance. Remember that observers don't have to be individual organisms; they can be neurons, regions of neurons, machines, circuits inside machines, and so on. The appearance and flow of information, sharpened and revealed as consistent patterns in a world of logical constraint … we are finally seeing how the knowledge of physical laws and the meaning of those laws, and even intelligence and consciousness, can arise out of the contributions of many individual observing subsystems communicating. The mysterious link between communication and meaning is something John Wheeler pondered in his it-from-bit paper. The laws of physics — which are expressions of consistent, predictable constraint in the world — sharpen with the accumulation of all of these observations and acts of communication.

No Aliens in This Universe

This takes us back to the Introduction of this book. I told a story in which scientists of the not-very-distant future reach full-on crisis mode, as they are unable to detect any unusual chemistry on any of millions of exoplanets analyzed from afar, something we will likely be able to do soon. Considering the richness of life on Earth, and how early in Earth's history life appeared, how can the rest of the universe appear to be sterile?

The answer is that the universe we see is exclusive to the techno-biological super-observer that we are. Our universe *is* the super-observer, and this means we are the only intelligence in our universe. Any other system or superorganism-type network of information that might develop in the multiverse of possibilities (Chapter 6) necessarily dwells in, and observes, its own universe.

Furthermore, any two universes are unobservable from each other. In other words, we will never meet aliens. I realize that's a fairly extraordinary claim.

Let us define an alien as anything that exhibits levels of complexity and information retention like that seen in Earthly life, but which does not descend from an ancestor in common with us.[7] Aliens are independent observers, which according to the stuff approach, may have come together in a manner similar to how Earthly life came together, out of atoms and molecules combining under the right conditions on some distant planet. However, in our informational scenario, a universe is a set of relations that build upon each other gradually, a single pathway down the Ultimate Plinko Box. Any other pathway is a different universe — if you or a particle detector observes an atomic nucleus decaying during a specified time period, that's a different universe from one in which the nucleus doesn't decay. The two universes are unobservable from each other: You can't check out what happened in the alternative course of events, any more than you can check out what would have happened in your life if you had studied architecture rather than business. This makes sense, because one Ultimate Plinko branch is not consistent with a different branch. All it takes is for one universe to go one way and another to go a different way, and there will be logical contradictions between the two worlds. We have never observed a logical contradiction in our universe, so there is little reason to believe we would ever observe one.

Discovering an independent system of informational relations, one that is in place prior to our discovering it, would be like discovering a detailed article on the *Reign of the Mantelopes* wiki (page 33) that had never been looked at before, and which had no causal association with the rest of the information in the wiki, but which is somehow perfectly consistent with all of

---

7 According to the conjecture known as panspermia, life around the galaxy (and perhaps beyond) is seeded from the same ultimate interstellar sources. In that case, other life forms would not meet this definition of "alien." Regardless, the possibility seems unlikely.

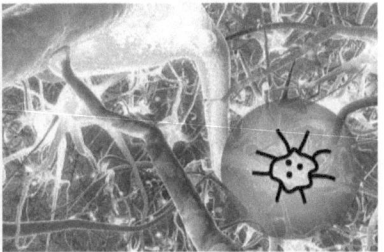

Fig. 7. A different observer system, disconnected from ours (e.g., an alien civilization), necessarily observes and dwells in an entirely different universe, perhaps with wildly different laws (right).

the other information that was gradually built up by fans asking questions about the *Mantelopes* world. If the universe reveals itself in the manner proposed by the simplest-case scenario, then there is no mechanism for a causally independent system of information to arise, as surely as there is no mechanism for a recipe to write itself on your computer hard drive. If the universe isn't *stuff*, why would an alien civilization observe our same Sun and Earth from afar, or see our spaceship approaching? There is no place in the simplest-case scenario for a detached system of information to arise independently, let alone one that is perfectly consistent with our own complex system.

For me, the only question here is whether the probability of finding a mutually consistent alien system is only vanishingly small,[8] or whether it's *exactly* zero, like the probability of finding another 2 on the number line. For astrobiologists and the SETI community, though, it makes no practical difference. In the 21st century, the rest of the universe will be shown to be devoid of any significant complexity, as described in the Introduction, and the simplest-case scenario will explain why.

I hope that in the decades and centuries to come, a new scientific paradigm will show that we are all in this together — not just figuratively, or socially, but also literally, physically,

8 There is one possible pathway to finding very simple alien life forms. The explanation is somewhat tedious, though, so I put it in the Appendix. However, the probability of finding alien *intelligence* is vanishingly small, if not exactly zero.

and mathematically. There may come a time when we realize that we are not disjointed, independent souls separately observing a world of lifeless stuff, but rather, we are a unified system, as unified as any individual organism fighting for its autonomy and survival. Perhaps only then, after a multi-generational sea-change of perspectives and priorities, will we finally achieve true collective enlightenment.

In *Cosmos*, Carl Sagan imagined that there might be an "Encyclopedia Galactica" cataloguing every civilization in the Milky Way. He highlighted one fictional civilization that had become so advanced, so enlightened, that it was virtually guaranteed to last for millions of years. The civilization called itself "We Who Became One." I like to think that the simplest-case scenario is the first step humanity can take toward the goal of becoming One.

Things to Remember From Chapter 8

• Informationally, the universe begins when life begins. The observation of things and objects prior to this event is an artifact of our modern description of the early universe.

• Life and the technology we build are unique in that they create informational relations that persist in the world, which are used to the biological or technological systems' advantage.

• All *techno-biological* observers are connected by a web of causation. This connectedness constrains them to observe the same logically consistent world. They also exchange information among each other, although this is limited by the speed of light.

• Any observer that isn't part of the connected techno-biological *superobserver* would observe and dwell in a different universe.

CHAPTER 9

THE EMERGENCE
OF INTELLIGENCE

I n a 2003 TED Talk, the physicist Lee Smolin discussed how our understanding of the universe has progressed through three historical phases. The first was the Aristotelian universe, which was based upon a hierarchy of celestial spheres. The second was the Newtonian universe, based upon a framework of absolute time and space. The third and current phase is the relational universe, initiated by Einstein in the early 20th century, and in some ways Galileo before him; see page 79. In the relational picture, there is no absolute time and space; instead, the world is seen as a network of interconnected relationships with no "outside." You could say that the simplest-case scenario is the result of taking the relational approach to the ultimate end. As I suggested in Chapter 5, it gives us a way to vanquish *all* absolutes from descriptions of the world. In the "stuff" picture of the nature of things, the information that identifies an electron by its charge and mass, for example, is assumed to be absolute. This information is entirely independent of other systems or observers, and it endures for a period of time — in the case of electrons, effectively forever. The assumption of absolute, independent, enduring particle-identifying information is necessary if the world is fundamentally made of stuff that's interacting according to laws. That's because if this information weren't absolute and enduring, then electrons would not all respond identically to electrostatic forces, for example — which they do, something that can be experimentally demonstrated. But in the simplest-case scenario, that information exists only in relation to the techno-biological superorganism and the subsystem that identified and measured it. Further, the universe operates on a need-to-know basis: Rather than some information being defined for all things in all places at all times, all information appears gradually, accumulating in the universe the way sugar crystallizes on a growing chunk of rock candy. When we conscious humans analyze this appearance of information in the world, we call the process observation. If there's no observation by some component of the superorganism, then there's no information, and that detail is simply not a part of the world. In this way, the universe is informationally minimal, while still accounting for every observation that has ever been made.

Consider, for example, a beam of light that we prepare using a laser. I claim, strictly speaking, that the beam does not consist of a stream of individual photons. How can I say this, when we know from physics that light consists of photons? Well, that's what makes the simplest-case scenario simple: The individual photons aren't individual features of the world, as long as no photon-information is extracted from the beam. If you aren't looking for the traces of individual photons, then those photons remain "unfocused and fuzzy," to recycle my favorite Paul Davies quote. Instead, the light beam exists only in terms of the minimal information associated with how it was prepared, such as the intensity and frequency selected, the radius, and the direction it's pointing in, relative to some particular reference frame.[1] All of these numbers were selected and/or known when we set up the experiment. But until we place a particle detector or photographic plate in the beam's path (*splat!*), not a single one of the individual photons is an ontic feature of the world. This explains how light can exhibit effects such as double-slit

---

1 Notice a similarity here to the life-as-a-river analogy on page 216: The information in the beam is not particle-specific and changing from moment to moment, but instead consists of broader, simpler parameters. The beam's overall intensity and the spatial relation between its boundaries are analogous to the flow volume and banks of the river.

interference, even seemingly retroactively in the case of the delayed-choice experiment (page 47). It turns out, we can't say that any individual photon went through one slit or another, because none of them individually was a part of the world at that juncture. In this way, much less information needs to be in the world in order to account for all behavior that the beam may exhibit. Nature does not need to specify every single photon, along with its information regarding frequency, spin value, etc. It is literally the simplest-case scenario for the light beam.

There's a pervasive concept in physics called the *principle of least action*, regarding the tendency of things to take the shortest or the straightest path. For instance, an ordinary object is said to always take the shortest path possible through spacetime — a straight line called a geodesic — but when spacetime itself is curved, the object's path appears to curve as well, which manifests as the effect of gravity. You could say that the simplest-case scenario extends this idea into a *principle of least information*. The world seems to prefer minimalist tendencies: Things take straight paths, objects in motion stay in motion, and laws of physics can often be unified into simpler versions, such as Galileo's discovery that all objects fall at the same acceleration. It makes sense that information should be minimal, too.

Speaking of the world being minimal, we now see how the universe (perhaps any possible universe) is built out of only one ingredient: the yes–no, on–off informational relation that we technological humans call the bit, or binary digit. Although we learn in school that the world is made of a variety of particles of matter and energy that interact, the simplest-case scenario explains this more complex picture as having emerged out of an arrangement of bits. Our universe's system of informational relations includes complex observing subsystems, some of which are very complex: humans and other highly intelligent beings.

The important things to grasp are that *everything in the world* consists of information interacting with other information, and that all of this information is contained within one system, the techno-biological superorganism. In the previous chapter I talked about how cellular and subcellular processes are information interacting with information, and that many technological devices mimic the way biological subsystems register, process, and store information. This generalization extends to every natural, mental, or technological process you can name. From protozoa in the guts of termites breaking down cellulose, to internet routers interacting with servers spread out around the world, to your central nervous system telling your heart how to contract, to Facebook groups discussing summer blockbuster movies, *everything* is information interacting with information.

If our universe evolved into its present state by gradually complexifying via the incorporation of new information on top of old information, we can identify the general stages that a universe like ours may evolve through.

The Stages of Universe Complexification

Stage 0: Periodic/cyclical system. This is a universe with no complexity, containing simple, regular relations. It never gets off the ground, evolutionarily speaking. A chain or loop of alternating relations is one such universe. The top of the diagram on page 159 shows universes that fit in this category.

Stage 1: Primordial system. This is a simple universe with rudimentary complexity, in the sense that it contains irregular or asymmetrical relations, like those shown at the bottom of the diagram on page 159. This kind of system evolves in a series of steps, with each step adding new relations that complexify the system. The arrow between these steps that points in the direction of new, additional relations (as in the diagram) is the emergence of a primordial arrow of time. A universe of this form may be considered to be one minimal observer — not to *contain* one observer, but to *consist* of only one simple observer. In our own universe, applying a modern description, this stage corresponds to the appearance of the earliest life form capable of registering and durably storing information in some respect.

Stage 2: Multiple-observer system. This contains different subsystems that can acquire information independently and evolve separately, but which remain causally linked or entangled (see Chapter 8). In our universe, this stage corresponds to multiple kinds of information-processing happening at once, either within one organism (i.e., it has more than one internal subsystem) or multiple simple organisms that have replicated. Regardless, the subsystems observe the same mutually consistent world, being part of the same closed system.

Stage 3: Communicating system. This contains different subsystems that can acquire information independently, not only from what is emerging to be their environment, but also from each other. In our modern-described universe, this might involve the beginning of coordinated cellular functions, or basic chemical sensitivity between multiple primitive organisms.

Stage 4: Diverse system. Not only are there multiple distinct subsystems, but they also have markedly different degrees of complexity, with bidirectional information flow (page 203). In our modern-described universe, this is where evolution begins to diversify the forms of life. There may be predation: Communication — broadly defined as the acquisition of information by one subsystem from another — may involve organisms observing the presence of other organisms, and then eating them. This also marks the emergence of true intentionality, i.e., organisms' freedom to move through and impose their causal efficacy on the emerging environment as well as on other organisms. The "Cambrian explosion" of 540 million years ago falls into this stage.

Stage 5: Proto-conscious system. This contains diverse subsystems that can store, process, and communicate information with advanced complexity, such as the capability to observe and compare internal states on multiple levels, and to create relations across time, this information-management naturally enhancing evolutionary fitness. In our universe, that may mean the evolution of animals that can communicate messages with specific content, such as some insects and fishes, and which can learn to some degree, and make decisions informed by past experiences.

Stage 6: Conscious system. Subsystems can store, process, and communicate and compare information at very advanced levels. They have the ability to create multiple internal states that reflect possible options to be acted upon, such as an animal imagining scenarios of what might happen if it chased prey that is borderline too large. They can respond differentially to information about individual subsystems, including themselves: recognizing kin, or learning that a reflection is not another individual, for example. Many birds and mammals, including pets, can do these kinds of things. (Yes, your cat is conscious!) Perhaps our universe reached this stage 80 million years ago, or something like that.

Stage 7: Symbolo-sentient system. This is the stage our universe is in today. Some subsystems — in our case, humans — have experienced a giant leap in information-processing and communicating capabilities, powered by the ability to turn information into *symbols*, such as language and writing, which become communicated not only to other individuals but also within themselves. These organisms can speak and think, in words, or some other equivalent symbolic means. Since they are sentient beings, they may be aware that the future is uncertain, or aware of their own mortality. They may ask analytical questions of "why," which simpler beings without language-like abilities and explicit self-awareness cannot. They plan and fashion tools to help accomplish those plans, which means technology — including the making of tools that make tools. Humans reached this stage only in the last 200,000 years or so, but other animals such as whales or dolphins may have a form of symbolo-sentience as well.

Stage 8: Techno-dominant system. In this ultimate stage, technology allows subsystems to be informationally connected virtually without limit. For the first time since Stage 1, the system regains an overt unity across all of its subsystems, as the inter-connectedness of everything in the system becomes known to its sentient inhabitants. Further, the ability of technology to store, process, and communicate information exceeds the abilities of the most complex natural systems. This is when technology is able to create its own superior technology. For us, the crossover to this stage (dubbed the "technological singularity") has yet to occur; authors and futurists predict it will happen in the early-middle part of this century. Assuming we are not overtaken by robot overlords, we will experience a full merging of biology and technology, with technology wired directly to our physiology. The glowing handheld devices that we presently gaze into — an interface that is awkward and dangerously distracting — will seem as antiquated as the clay tablet. Biological and technological information management will become seamlessly integrated and astonishingly powerful, perhaps with minuscule information-registering and processing subsystems ("nanobots") coursing through our bloodstream and maintaining our bodies as no present medical technology can.

Now, to clarify what I mean by these classifications, if a universe is described as being at Stage 7, that's a description relative to (or as seen from the perspective of) its most complex subsystems. This whole book has been about vanquishing abso-lutes from descriptions of the world, and relative to simpler organisms, our own Stage 7 universe would have to be described as being in earlier stages. To a bacterium or a plant, for example, the universe appears to be at Stage 4; to a dog, it seems to be at Stage 6. The world is a very different place for a moth circling a streetlight than it is for us, as we watch and write down linguistic symbols about a moth circling a streetlight. What does an ant know about the computer it's crawling into? An ant's world is largely about quantized information that we call chemical cues, organic molecules. From the perspective of an individual receptor on an ant's antenna — its processes making up a proper subsystem in its own right — this might be a Stage 3 universe.

I don't make this point breezily or to be cute. The completely relative nature of the simplest-case scenario requires that descriptions are not absolute; they are relative to the subsystem having the information in question. Recalling that descriptions are packets of information contextualized and enriched by other information (see pages 169–170), if a subsystem does not have much contextual information or information-synthesizing abilities at its disposal, its experience of the world will be accordingly simpler. Of course, we humans are the ones making this extremely rich description of a description, in symbols that we call words (language is discussed on page 258). That's the kind of complex information-building we symbolo-sentient subsystems do, and it's why the world is so rich and detailed, *relative to us.*

One more thing: I referred to Stage 6 systems as being conscious. Invoking a "conscious universe" is like something you might hear from Deepak Chopra, who has many critics in the physics community. But if we view the universe as a techno-biological system of information, and the system contains conscious observers, then it's natural to think of the whole connected system as conscious. After all, you would probably consider yourself conscious, as a person in a human body, even though there are many parts of you that exhibit little or no consciousness, such as your teeth. Each part is a part of you, as a whole. Notice, again, the fractal nature of this perspective: Your consciousness is a feature of your own diverse personal subsystem, just as the universe's consciousness is a feature of the entire diverse system. The simplest-case scenario operates the same at every level and scale, and so the world can be viewed similarly at every level and scale.

How Behavior Emerges From the Genome

One of the most spectacular mysteries of life is how the complexity of a living organism, including all of its physiology and instinctual behavior, can ultimately be traced to the base-four code of its genome. This is a perfect example of emergence: It starts with the finite, compact informational DNA sequence, which is nothing like a representational map of the body. Even though every detail of every part is not specified, as in a blueprint, cells are able to differentiate and to coordinate with other cells of other kinds, to form organs. Higher functions emerge from these interactions as well, such as the coordination of organ systems, the transfer of non-genetic information between parts of the organism (hormonal and nervous signaling, etc.); the list goes on and on. The emergence of instinctual behavior, innate responses that are not learned, is just complex enough to be thoroughly astonishing, and to warrant a close look.

I once had a mosquito larva in a bowl of water. I wanted to see what would happen if I added a bit of soap. Mosquito larvae breathe by piercing the water's surface tension with a tube from their tail, and soap destroys the surface tension. What would happen if there were no surface tension? What I saw amazed me: With the larva's breathing tube clogged with water, the insect bent its abdomen and nibbled at the tube in an attempt to unclog it with its mouthparts. This is an instinctual routine — a contingency plan in the event that the breathing tube becomes obstructed — that every larva apparently carries with it, thanks to natural selection. The behavioral routine has evolved, likely out of several related but simpler routines, such as the defense mechanism of contracting to evade predation. By the time present-day mosquitoes had evolved, individuals without this complex survival routine had been weeded out. This all leads up to the complex behavior that we can witness today on the kitchen table: It appears that when a certain stimulus threshold is reached, perhaps a drop in cellular oxygen levels, a mosquito larva's abdominal muscles contract, and the minuscule muscles of the mouthparts differentially contract to perform highly specific movements in relation to the tube. Incredibly, this complex behavior somehow emerges out of the DNA code! How? It astonishes to the point of seeming impossible.

This is a case where an informational approach can be enlightening, because it deals directly with what DNA is in the first place. The mosquito appears to have raw genetic information going into some "black box" that we don't understand, with observable complex behavior coming out. Recall our discussion in Chapter 5 of information sharpening or constraining other information, due to the action of context. The black box in this case is all about contextualizing the genetic code, even while this contextualization comes *from* the genetic code.

Imagine that someone in the 1930s was handed a printout listing all of the 1's and 0's that would be found in a modern digital format of Bing Crosby singing "Brother, Can You Spare a Dime?" Then imagine that we asked the person to hum a few bars of what they were looking at, without telling them what it was. Naturally, they would be flummoxed. If you fed the pages into a box that could scan and play the recording, and then you asked the person how this piece of music can emerge from a list of numbers, they would be equally clueless. The numbers go into a box, *something* happens, and music comes out. But this cluelessness and astonishment result from not understanding the contextual goings-on in the intermediate step, the decoder box. This is what a person would need to know in order for the process to become clear: (1) The 1's and 0's come in chunks of 16. (2) Each chunk of 16 bits is a binary representation of a familiar base-ten number; for example, 0010100111100011 is equivalent to the base-ten number 10,723. (3) Each of these base-ten numbers represents a level, which can take any of 65,536 values — that's how many different combinations that 16 bits can be arranged in. (4) If you were to put 44,100 of these level values together on a graph, and then you connected the dots, you'd get the shape of the audio waveform for one second of music. In fact, the shape of that wave would resemble the physical shape of the groove on an old record of "Brother Can You Spare a Dime?" — something that your friend from the 1930s could verify with a magnifying glass.

Consider how much contextual information has to be known before it becomes clear how a song can emerge from a list of 1's and 0's. We needed to know how the string of bits was divided into larger units (by chunks of 16). We needed to know what each of those units represents (a level value). We needed to know the time interval between each of those values (1/44,100th of a second), and we needed to know what to do with the level values separated by the time interval (connect the dots). Given that context, even someone in the 1930s with graph paper — and a lot of time and patience — could draw an approximation of the waveform for the song, or any part of it. In fact, if you could transform that drawing into a carved groove in wax or acetate, the string of bits could be played back in audio form.[2]

Bear in mind that this example of converting numbers into sound is a basic technological process. Most biological processes, even at the cellular level, are much, much more complex. Unlike the linear logic of digital audio, biology is not intelligently designed. The difference between manmade, intelligent technological encoding and the organically evolved genetic encoding of biology is like the difference between the logical, systematic layout of the streets of midtown Manhattan versus the unplanned convolution of Venice. You need a lot more context — *knowledge* — to navigate in Venice than you need in Manhattan. But, if you were privy to all of the physiological messiness involved in the transcription of mosquito DNA and RNA into proteins, and the processes that arrange those proteins across the various differentiated cells in the insect's body, and the manner in which the organism senses, stores, and compares stimulus information, and so on and so on — then, in principle, you *could* chart out, from beginning to end, how a set of genes leads to the emergence of this amazing survival behavior.

---

2 The CD-quality standard of 16 bits and 44,100 samples (levels) per second — in stereo, with separate left and right channels — would be overkill for encoding recordings from the 1930s. Our Depression-era listener might be satisfied with a monophonic recording using eight bits and 8,000 samples per second.

What's even more amazing, though, is that in biology, such contextual information *also* emerges from the genetic code. It would be as if the digital code for a song, interspersed with the bits representing the music, also embedded instructions on how to build a music player from raw materials — including how to build every tool needed to build one! We see yet another fractal in life: An organism builds itself out its own old (genetic) information, with the added influence of new information from the environment, to which its lineage gradually evolves to adapt. This process is similar to the universe building itself based upon its own legacy information, plus the added influence of new information appearing in the world, resulting in the emergent laws gradually sharpening, to the point of the universe seeming to be fine-tuned for life.

A similar process goes on even when a human being designs something, like a watch: A watchmaker starts with old legacy ideas on how a watch should be built, and adds new ideas in order to fine-tune the watch's accuracy. For millennia, a great many people have noticed these parallels between nature, particularly life, and objects that are designed by humans. But in concluding that life and the universe must be intelligently designed, these people have been looking at things from the wrong perspective: Actually, human invention and design are purposeful, sentience-directed variations on the same informational processes that occur naturally in the evolution of life and the universe.

From the instinctual behavior of mosquitoes, things get considerably more complex. In mammals, for example, behavior such as nesting and parental nurturing is nuanced and flexible and dependent upon equally complex environmental factors. Learned behavior is in yet another league of complexity. Then of course there are humans, with our highly specialized adaptation behavior as well as formal social behaviors, such as ritual. We even have *non-adaptive* behavior that emerges from adaptive behavior! The ability to create and enjoy music is not something that was evolutionarily selected for, but it emerges from other emergent behavior, such as the ability to perceive, memorize, and recall sounds, as well as aesthetic faculties, which are probably rooted in sexual selection, the evolutionary influence that's responsible for the peacock's tail. All of these complexities build from lower complexities. The point is, if we knew and understood *all* of the contextual constraints involved, we could, in principle, work out how even the highest and most advanced human behavior emerges from the lowest: the genetic code, plus information from the environment.[3]

When we talk about the genetic code, we aren't really talking about the emergent chemical units that make up DNA, such as cytosine and guanine. That's *stuff*. The level at the bottom that we're interested in is the information in the code — the arrangement of informational relations from which those four chemical nucleotides emerge, relative to humans, with our knowledge of chemistry and our laboratory equipment. But the nucleotides are just the beginning. From the informational arrangement of genes ultimately emerges the way those relations relate with other genetic relations, and the way those genetic relations relate with other relations that have emerged from the genes, and the way emergent relations relate to each other *and* with environmental or new information, as in the example of a mosquito-larva cell comparing oxygen values across time.

All of this is a direct parallel with John Wheeler's suggestion that when we talk about the fundamental makeup of the universe, we should be talking about information. The chemicals that we call cytosine and guanine emerge to us when we look closely at DNA as a polymer of matter, just as particles and fields and forces emerge when we query nature in other ways to high precision. And yet, informational relations (bits) are really all there is at the very bottom.

---

3 Information from the environment includes the epigenome, which switches genes on or off. Epigenomic effects, a fairly recent discovery of great importance, are the responsibility of proteins on chromosomes, which are in close contact with the DNA.

Reproduction, Knowledge & the Scientific Method

Reproduction or birth is one of the most profound transitions we humans can experience. Indeed it is amazing that two people can come together, and nine months later, there are three people. It's all the more amazing given the incredible complexity of life and physiology, down to the subcellular level, such as the digital switching and informationally orthogonal processes discussed in Chapter 8. Yet under ordinary circumstances, reproduction is extremely reliable and works without a hitch.

In the simplest-case scenario, the reproduction of individual organisms is a special case of a much broader understanding of reproduction. Defined generally, reproduction is any copying of information: We start with one relation or subsystem of relations (however it might be delineated), and we end up with more than one relation or subsystem. Some examples of processes that qualify as reproduction, according to this broad definition:

1. A man and a woman have a baby.[4]
2. An aphid or komodo dragon produces offspring asexually through the process known as parthenogenesis.
3. One multicellular organism (such as a hydra) divides asexually into two organisms.
4. One cell in a multicellular organism divides into two cells.
5. DNA unzips in half, and each half is re-completed by binding the appropriate nucleotides.
6. Messenger RNA (mRNA) is synthesized from DNA.
7. Units of transfer RNA (tRNA) bind to mRNA.
8. Protein is synthesized from tRNA.
9. Someone takes a sequence of 1's and 0's, and constructs a corresponding sequence where every 1 has been replaced with a 0, and vice-versa.
10. A physicist hears an associate say, "I measured the spin of the particle along the up–down axis, and found it to be spin-up."

---

4 Bearing in mind that the delineation of a subsystem is arbitrary, both parents can be considered as being a single subsystem for the purposes of this example.

11. I get an idea and I write it down.
12. I click SEND on an e-mail addressed to someone.
13. Your phone receives an automatically generated alert message from your bank.
14. The IP address for a new internet domain name propagates to DNS servers on the web.
15. You read a word in a book.
16. I type the number 42 into a desk calculator.
17. I hear my smoke alarm go off.
18. My DVR records a television program.
19. Someone sees light from a flashlight that a friend has just switched on.
20. A man takes a swing and punches his enemy.

Some of these look like acts of reproduction, and others do not. But if we consider that everything is ultimately information, and reproduction is any copying of information, then we see what all of these processes have in common. Any kind of communication is reproduction, because it's a process of transferring a copy of some information from a sender to a receiver. That includes any writing down, recording, or reading, whether it's done by a human or a machine. Also, any kind of causal action on the part of one subsystem, which affects another subsystem, is a form of communication: If a man winds up his fist and imparts momentum into it, his victim will shortly thereafter receive this momentum information, probably rather painfully. Switching on a flashlight sends information, at the speed of light, about the new configuration of a circuit where a battery is wired to a lamp. One technological device communicating with another is also information copying and is therefore reproduction.

The subtleties at work here are where things get interesting. The copying of information isn't always perfect; mistakes can be made, generally depending on the complexity of both the subsystems in question and the information involved. A desk calculator, with its numerical interface and relatively simple circuitry dealing with definite numerical values, is extremely unlikely to make a calculation error. But Siri — or even more so, Watson, the IBM computer that competed on the game show *Jeopardy!* — has to interpret input, deal with large databases, perform multiple simultaneous data-juggling routines, and navigate decision trees. These information processors are much more likely to misapply input and supply output that we would call incorrect. In a similar way, a bunch of humans, sharing extremely complex and nuanced information via an imperfect language system, are more likely to generate and adopt untruths than, say, a group of birds communicating about local predators (although that's certainly not immune to error, either).

This kind of information-interpretation and copying results in there being something that we call knowledge, a variety of information that's not necessarily the same as the information defining the reality of the world. Suppose a primitive animal, attached to a rock in a tidepool (I will use a modern description here), physically contracts in response to changes in light, so as to defend itself from predators. It may respond when any organism swims by, whether it's a predator or not. It may even respond to the sway of a leaf. Trying to imagine the world from the perspective of this organism, it senses changes in light. This change is an informational feature of the world, which happens to have been caused by a passing organism; the receipt of this information is an act of reproduction, as defined broadly above. However, if it's responding to a change caused by a non-predator, or by a leaf, you could say that the organism is misinterpreting the information. Of course this animal isn't actually thinking that a predator might be approaching — it's just an autonomic response — but the process of natural selection has created this behavior, imperfect though it may be, simply because an organism without this defense mechanism is less likely to pass its genetic information forward. As a result, sometimes the response is a false positive, occurring even with stimuli that have nothing to do with the response's evolutionary origins.

Now consider a higher animal, such as a mammal. If a deer hears a branch falling in a forest, that may evoke an image of a possible predator as the animal turns its head. This is not an accurate reflection of reality, but rather, it's a misinterpretation of the sound-information received. The mammal may weigh the risks and benefits of taking various possible actions, and then maybe run, due to its interpretation differing from reality. It's in the deer's best interests to err on the side of caution — better safe than sorry — so here is a case where misinterpretation is encouraged by natural selection.

With humans, there are innumerable ways that information can be misinterpreted or miscopied. Someone may hear a phone number and write down the wrong digits. Even though the original number is correct, in mishearing and then writing down this information, the person acquired false knowledge. And that's a simple case: When dealing with ideas and abstract propositions that can be communicated only via the fuzzy form of information transfer that we call language, the possibilities for miscopying and misunderstanding are legion. A person's or a group's knowledge and view of the world can drift further and further from reality. I'm sure you've seen this happen — maybe in your own family!

This doesn't signal a defect in the world. The logic of the universe is always 100% consistent, with exactly zero contradictions between old bits and new bits that appear. Experiment after experiment has shown this, and indeed it's why we can write down laws of nature in the first place. The information behind the logic of the universe, which constrain observations system-wide in a spooky-action-at-a-distance manner, is what makes up reality; those are the bits John Wheeler wrote about, the bedrock of the universe, from which stuff, space, and time emerges. However, such self-consistency is not necessarily true of information that one subsystem acquires from another, in acts of reproduction, broadly defined. Knowledge falls into the latter category, and sometimes, knowledge contradicts reality.

Which brings us to the endeavor called science. What is science, if not the human effort to reconcile the discrepancies between reality and knowledge? Like the organism that contracts even if no predator is passing by, which is a primitive kind of false belief, humans misinterpreting information leads to false knowledge and untrue beliefs, only on much grander scales. Science, then, is the concerted effort to learn where misinterpretation has led us astray, and to bring these two informational streams — reality and knowledge — back into alignment.

Science accomplishes this in a remarkably effective way: First, by simplifying and removing complicating or unpredictable influences, so that specific effects can be isolated and examined. Second, by describing individual aspects of nature mathematically, in equations that express their relations — not only so we can better understand things that have already happened, but also to reduce our uncertainty of the future. Starting with predictions of eclipses in ancient times, science has successfully been able to generate information about the future, even if that information isn't coming *from* the future! Third, science uses technology to improve the information-registering and interpreting process. A photon detector/counter is more reliable and objective than a guy with goggles saying, "I think I saw something!" Finally, in science, many people and multiple technologies check both the experimental data and the interpretation of that data. This process, performed over thousands of years, has given us knowledge of ever-closer approximations of the reality of the world. We have come from being helpless to the smiting of capricious gods, all the way to measuring an obscure asteroid's motion in space and predicting where it will wind up years later. Meanwhile, our medical knowledge has progressed from evil spirits possessing us, to imbalanced "humors," to the awareness of disease-causing microbes, and consequently, how best to deal with them.

If the simplest-case scenario is an accurate account of the bottom layer of reality, then we won't ever fully understand how physical reality comes together by continuing the assumption of stuff as fundamental. Science probes matter on progressively smaller scales by colliding particles at higher and higher energies. This has produced the Standard Model, where for example a proton is known to be composed of three quarks. Perhaps by achieving higher collider energies someday, we may find a deeper level of the Standard Model, that quarks are composed of sub-quarks or something. Probing subatomic particles to greater and greater precision may generate more information about stuff and its properties, without necessarily getting us closer to answering the biggest questions, such as, "Why these laws?"

Studying stuff by probing it at higher and higher energies is a bit like calculating the number π to great precision (mentioned on page 163). Calculating π to a billion decimals generates a lot of digits, but that string of digits doesn't tell us anything about what π actually represents or what it means. If we didn't know that π represents the relation between a circle's measured cir-cumference and its diameter,[5] calculating those digits would get us no closer to understanding. Tens of thousands of digits, by themselves, are more like stuff: The information goes on and on without end. If you want more, just calculate some more. You could fill the universe with calculations and you still wouldn't get to the bottom of π — just like you could fill the universe with observations, generating a huge quantity of information about stuff, and you wouldn't get to the bottom of where it all came from or why the laws of physics are the way that they are.

It's a very curious feature of the universe, these two ways that information can behave. On the one hand, the physical–factual constraint generated by the system's acquisition of information is consistent and contradiction-free. But, when we reproduce information from one subsystem to another, that process is fallible. Also, new information can appear that recontextualizes

---

5 As a constant, π appears in many physics equations that aren't concerned with length measurements of circles, and it can be calculated in various ways that don't involve circles or length measurements.

(reframes) or even changes old information. In genetics, such changes are known as mutations, which sometimes happen when a DNA replication error is not corrected. But life is at the mercy of other disruptive effects, as well. For example, mutations can be caused by radiation or high-energy matter such as beta particles. Why are these monkey-wrenches routinely thrown at us — random events that change the course of affairs in the world, from a gamma ray that causes one mutation in one cell, to a meteor that wipes out most life on Earth?

It turns out, despite "fine-tuning," the world is surprisingly hostile to life; mutations and meteor impacts happen! By inspecting nature closely, we have learned to understand these things in terms of stuff: matter and energy following the laws of physics. In the case of beta radiation ionizing a nucleotide, the properties of beta particles are in accord with the Standard Model that we have formulated from our many other discoveries about matter and energy. In the case of an impact leading to the demise of the dinosaurs, we have discovered that outer space is filled with rocks and chunks of ice that occasionally hurdle our way. But recalling the anthropic principle (page 143), apparently it *has* to be that way in our universe — mutations are critical to variation between individuals, and variation is critical to natural selection. If there were no informational variation, we never would have evolved beyond single-celled organisms, let alone to the point of intelligence. And, if it weren't for an impact 66 million years ago, a reptile might be sitting where you are. In our universe, and perhaps all universes where advanced hyper-complexity can evolve, there is an element of chaos, of wild turns of events due to unexpected new information as well as reproduction-related alterations (imperfect copying). Without these monkey-wrenches thrown into the natural course of events, we would not be here.[6]

---

6 Incredibly, The Epicurean philosophy of Ancient Greece recognized this. In his tribute to Epicureanism, *On the Nature of Things*, the Roman poet Lucretius wrote of a force that randomly caused minute particles to change course or "swerve." Were it not for these swerves, the world would be perfectly ordered and uniform, for all of eternity.

We are learning more about the importance of chaos to the emergence of complexity. Using simple interacting digital models called cellular automata, computer scientist Christopher Langton studied the evolution of patterns in systems ranging from ordered to chaotic. He discovered that in very ordered systems, there are patterns, but none of them develops any complexity; there is just too much regularity for anything to get going. In very chaotic systems, patterns may begin to develop, but the probability is low that any pattern persists for long. However, in between the ordered systems and the chaotic systems is a "Goldilocks zone" that's just right for complex patterns to appear, persist, and evolve. This suggests that in a similar fashion, considering the set of all possible universes (Chapter 6), universes may be ordered or chaotic — but for a universe to develop the kind of complexity observed in our world, it needs to be somewhere in between. And, indeed, life does seem to be balancing on the edge. If life had been disrupted less, it might have evolved slower, or not at all; if life were heavily challenged by a very chaotic state of affairs, it might have gone extinct. You could say that this balanced position is "fine tuned." Regardless, the imperfect struggle of biological information to maintain and copy itself, amidst moderately chaotic and disruptive effects, seems to have supplied the impetus for life's own growing complexity and, eventually, its own conscious, sentient self-discovery.

Fig. 1. In cellular automata that are very regular (left) or very chaotic (right), diverse patterns do not develop. Complexity is most likely to appear in intermediate types (center). In these one-dimensional cellular automata, evolution begins at the bottom of the diagrams and progresses upward.

Consciousness: Follow the Information

Is there an immaterial element to ourselves, like a soul, that perhaps can pass into another plane of existence upon death? We'd like to believe so, because most of us don't want to die. But there is no evidence that anything like a soul survives death. Science has helped us to better understand what death means, but people believe regardless. This speaks to various things: the difficulty and emotionality of this transition, ingrained religious teachings, and the wants and desires of humans. But ultimately, these beliefs arise out of an insufficient understanding of the relationship between the mind and the body, often called the *mind–body problem*. Can the mind and consciousness be reduced to the atoms and molecules of the brain? Some physicists and prominent philosophers, such as Henry Stapp and Thomas Nagel, have doubted this in recent years. Their work has quite naturally been met with skepticism, but they have reached these positions out of sheer desperation, all other options seemingly having been exhausted.

Reducing things to stuff such as molecules, and then atoms, and then subatomic particles, which obey physical laws, is a fine way to model the behavior of much of the world: galaxies, stars, and planets, as well as smaller objects. With the assistance of chaos theory and computers, even notoriously complex systems such as Earth's atmosphere can be modeled and predicted to some degree. But as physicists and philosophers have written, strict reductionism does not account for all of the nuances that are intrinsic to living biological systems. A good example is bidirectional causation — both reductionism-friendly bottom-up causation and top-down causation, acting in coordination and at multiple levels. Bidirectional causation, and top-down causation in particular, is more comprehensible when viewed from a fundamentally informational perspective: The higher a function is, the more obvious the role information plays, with the conscious mind being the ultimate case of a high-level process capable of affecting lower-level processes. When you think about an upcoming event that makes you nervous, there's an increase in the contraction rate of heart-muscle fibers, which then affects your general physiology, which affects the mind. Complex interacting feedback loops such as these, both positive feedback (which amplifies effects) and negative feedback (which buffers and stabilizes effects), are distinguishing characteristics of life and most definitely of the mind and body. Perhaps information theory can help us tease apart the mind and the body.

So what exactly is *a mind*? The informational perspective tells us that a mind is a complex subsystem consisting of many simpler interacting subsystems, all of which ultimately consist of informational relations, like everything in the world. This is not to be confused with *a brain*, which is the emergent material substrate for the mind (more on that in a moment). The brain is different from the mind a bit like the way a refrigerator is different from refrigeration: A broken refrigerator may have all of the same parts as a working refrigerator, but for whatever reason, it cannot refrigerate. Perhaps a wire or a plastic piece has become disconnected, interrupting one part's ability to electrically or mechanically link to another — literally a break in the causal chain. Similarly, a functioning live brain and a dead brain may have essentially all of the same atoms, but the living brain exhibits the properties of *mind*, while the other does not. The material infrastructure of a dead brain may still be there, for the time being, anyway, even if the complex and bidirectional causality flows aren't.

It's well understood that while conscious functions such as thoughts are certainly a component of the mind, they are only a part. There are many layers to the mind: conscious processes, subconscious processes, memory faculties that operate in the background, and routine or autonomic processes, such as the regulation of breathing. There are also many *levels* of complexity to the mind, from high-level functions such as reasoning and intellectual thought, to the lowest, such as individual molecular-scale processes that we have come to associate with the biochemistry of neurons. The mind comprises all of these functions. Then again, who's to say the mind should be limited to what goes on inside the confines of the cranium? Our body's information superhighway, the nervous system, is an undivided structure that pervades the entire organism. Information flows on "surface streets" as well, such as the bloodstream, where hormonal levels fluctuate and are affected by the nervous system — and which, in turn, affect the nervous system: bidirectional causation in action again. So, you could argue that the mind extends in space to every cell that participates in this incredibly intricate dance.

Even *that* organism-level boundary becomes somewhat arbitrary, though. Anytime one person exchanges information with another person, they achieve a certain degree of mind-meld. Taken to the extreme, then, the idea of the mind extends ultimately to all humans and indeed all living things — the entire universe, in this highly biocentric way of looking at things. Such mystical or Eastern-sounding pronouncements are not a part of mainstream science. But saying that the universe in some sense can be considered *a mind* is an unavoidable consequence of (1) John Wheeler's it-from-bit declaration that the world is fundamentally made of information, (2) the conjecture that this information is limited to the techno-biological superorganism, with all the stuff of the world emerging to observers within that system, and (3) a consistent usage of terms. The universe is, in some sense at least, *a mind*.

You might argue that all of this is pointless semantics, and in some ways, you'd be right. I personally have no problem with saying that the mind ends at the individual person. But, how do we define a mind in the case of twins conjoined at the head? Or in a person whose left and right brain hemispheres have been surgically separated? At least now we have a way to be intellectually honest that *a mind* is not necessarily a one-size-fits-all term. We have a way to acknowledge the gray areas and nuances. As with concepts such as species and gender, in most cases the meaning of the word "mind" is clear. But it need not be absolute.

Let's step back a bit and talk about conscious experience, something we can relate to more easily than molecular processes; we're living consciously right now. Conscious experience is a subjective phenomenon, exclusive to the individual experiencing it, that involves all of these processes of the mind that I've been discussing. We get new information through our senses, and our mind places that information in relation with various relations between old information, perhaps resulting in a causal action that reaches beyond our own individual self. Here's a real-world example: Suppose someone calls me a jerk. That message is information, since it's capable of reducing my uncertainty about what this person thinks of me. I register that information in raw form, which in the stuff-oriented description means sound waves creating neural impulses from my inner ear. From there, my mind will interpret the impulses with the help of learned associations regarding language — old, legacy information — which provide context for the raw information about sound. (Recall the parallels with the SOS story from Chapter 5.) This context allows me, as an individual with a conscious mind, to extract meaning from the neural impulses: I now have an informationally enriched version of the sounds that reached my ear, which reduce my uncertainty far more than the raw sounds alone would. This meaning having been decoded — my adversary originally encoded the meaning into sound when he uttered the insult, using similar legacy information about language — I may recall a memory from the 5th grade, when a bully called me a name and made me wet my pants. (That didn't actually happen, just so you know.) This recollection might trigger emotions of embarrassment and anger, which then in a top-down manner causes my heart rate and other physiological details to change. Finally, with the help of my language faculties and memory bank of grade-school retorts, I may respond verbally, another top-down influence in which my diaphragm contracts, my vocal cords tighten, and my mouth shapes the perfect comeback: "I know you are, but what am I?"

That is a very broad account of what's going on in that situation. It simplifies or ignores the many subroutines that occur at deeper levels. The recollection of a distant memory, for example, is an enormously complex process in itself, and poorly understood. In the decades to come, though, these kinds of mysteries will likely only be solved by looking at the mind in terms of the information it holds, and how that information got there.

Language: The Rocket-Engine for Humans

How did great apes, in only a few hundred thousand years — a blip in the evolutionary time scale — suddenly take off and become so intelligent and successful? As far as we know, at no other time in the history of life has a genus advanced so far and so fast, to the point of developing technology, for example. The onset of language abilities is a popular focus for theorists. While it's certainly possible that language emerged from other evolutionary developments, the informational perspective suggests that language actively drove the evolution of mankind.

What's so extraordinary about language? Throughout this book, I've argued that information can be sharpened through the application of context. As it happens, spoken language is a systematized, standardized method to sharpen information with contextual framing, so as to transmit highly meaningful messages and descriptions. Meanwhile, written language is a method to record this information durably, the way the genetic code is recorded durably. In a world fundamentally made of stuff, these advances would be greatly advantageous — but in a world of information, they are evolutionary sea-changes.

Other species have ways to transmit sharpened information; the waggle dance of the honeybee (page 114) is an example, where a message encodes not only the yes/no answer of whether there is food nearby, but also its direction relative to the Sun and an idea of distance. But think of the detail that can be compressed

For any information-processing action performed by the mind, however, there is an equivalent stuff-based account, a kind of ultra-modern description that would be reached if you could fill the brain with observations of the various associated particles — if you went in there with the appropriate technology and found them. The really fascinating thing about the brain is that if you look under a powerful microscope, you can see its neurons

into a sentence made of words. You can identify things (nouns), modify these identifications (adjectives and such), invoke action and color that action (verbs and adverbs), specify direction and location (prepositional phrases), and even express emotion (interjections). Inject further context with tense, tone, and so forth, hang all of this on a connective scaffolding of conjunctions and articles, and build paragraphs and larger structures that represent a coherent flow of ideas — and you get an idea of how humans took off like a rocket. Suddenly, very sharp information could be passed among a social species and quickly propagated among groups. As language complexified and vocabulary grew, it became possible to synthesize increasingly abstract and nuanced ideas. Without these advances, where would we be? No complex cooperation, no philosophy, no science, no written records of our discoveries, no true civilization.

Consider this: Even if a social primate envisioned a scenario where competing groups' territories were divided fairly into equal-size regions, this concept could not be clearly articulated without language. Other individuals could not be recruited to solve this and other problems and advance the larger group forward. (Perhaps this was the disadvantage of Neanderthals and other early hominids.) But we humans found a way to do that, and modern humanity — with societal organization, rapid accumulation of knowledge, and so on — is the result.

branching and reaching over relatively long distances to connect with other neurons, to facilitate the kinds of information exchanges that happen in consciousness. The world being a logically consistent place in which magic doesn't occur, and matter being something that has emerged to us, we necessarily *must* find a matter-based substrate — a material infrastructure of neurons — for the kinds of information flows that occur in consciousness. It's a tiny bit like having directions to get to a party, and then discovering that when you get out there and you look, you find actual roads that correspond to those directions. However, imagine if we had to figure out the brain only by looking at the individual neurons and attempting to follow and interpret the connections. The challenge would be virtually insurmountable, like looking at a genome sequence and trying to infer, only from the letters, what is being encoded. Gregor Mendel didn't study genetics by starting with the chemical structures of DNA (they hadn't yet been discovered); instead, he followed the heritable traits that pass from parent to offspring. Then, when the structure of DNA was later discovered, everything we knew about heritability from Mendel and his successors fell into place. Similarly, if we strengthen our understanding of the information flows and faculties involved in consciousness, this will surely facilitate the study of the brain's material wiring. We already have good clues as to where various faculties are distributed around the brain. Combining this knowledge with a strong theoretical framework on the informational structure of consciousness could lead to rapid advances in brain science.

Everything Real Is Imagined

There is one fact about consciousness that people nearly always neglect, and it's true with or without the simplest-case scenario: *Everything that you have ever seen, heard, touched, smelled, or tasted, is something that you imagined.* When you see an object, the internally experienced image of that object is formed exclusively in your mind. Your consciousness does not have direct access to the "outside world," whatever that might mean. The retinas of your eyes and your optic nerves and whatnot are very much involved, of course. But as Robert Lanza points out, they have to send their neural impulses through holes in your skull, and on the other side of that bony barrier, your mind puts the pieces together, like the telegraph operator in Chapter 5 putting together dots and dashes and extracting meaning from them.

As a result, the image of a very real object ends up in the same place, more or less, as an image of something that isn't "out there" but which you mentally create or dream. Both kinds of images are formed within the mind. It's as if the information for objects-in-the-world and the information representing imagined or dreamed objects winds up going through the same channels — and, sometimes, it isn't a trivial matter to tease these streams apart. Some people do not like to hear this, because they want to believe that they have an absolute handle on reality and a direct access to the outside world, but the truth is, nobody does! The mind is fallible in its ability to distinguish "outside" information from "inside" information, as a natural result of imperfect copying and interpreting by the faculties of the mind. This flaw is an inevitable part of the human and animal condition. When someone says, "I know what I saw," and they describe having witnessed an object disappear or perform some other magical act, well, that's an example of false knowledge resulting from imperfect information-processing. This is nothing to be ashamed of, especially when you realize that everything you see, whether real-in-the-world or only-in-your-head, is processed in a similar fashion. The only difference is how the information got there, and sometimes you can't tell.

An information-based treatment of consciousness, the admission that all images we experience are similarly mental, explains a wide variety of subjective phenomena. An example is ghosts. Steeped as we are in the stuff-based approach to existence, when someone asks a question like, "Do you believe in ghosts?" the assumption is they're asking whether we believe that ghosts are features of objective reality out there in the world. For anyone who has seen a ghost and knows what they saw, all I can say, as a non-ghost-seer, is that such an experience must be extremely convincing. People have been having these and other shockingly convincing subjective experiences for centuries: Lucid dreams, alien abductions, visions of the Virgin Mary, and mental-illness or drug-induced hallucinations are all cases where the line between objective imagery and mental creations becomes greatly blurred. The mind is a poor judge of the objectivity of its own images, just as it can be a poor judge of things like line lengths (page 15). This is why science places a high value on the use of technological instruments, peer checking, and reproducibility: These are all ways that collectively, we can align knowledge as closely as possible with objective reality.

There is an objective and measurable distinction to be made between these two kinds of conscious experiences, though. If I dream or hallucinate that I've measured a particle as spin-up, that action does not constrain you in any manner at all, any more than imagining you weighing 300 pounds constrains the behavior of your bathroom scale. So, there is a simple test for objective reality: The next time you see a ghost, be sure to have a Stern–Gerlach apparatus handy, and ask the ghost to measure the spin of, say, a few dozen particles. If those measurements verifiably constrain your measurements of the same particles in every case, then congratulations: You can now claim, with extremely high confidence, that the ghost was a feature of objective reality!

The Nature of Free Will

Free will is one of those seemingly intractable issues in physics, philosophy, and brain science. Do we actually have the freedom to make a decision, one way or another? Do we create those decisions in real time? It certainly feels like we are making a valid choice, but physics gives us reasons to doubt this feeling. According to some theorists, the universe may be deterministic, with all future events fixed and unchangeable, constrained in full by the dynamics of the universe in the past. If so, when we make a decision, our freedom to pick one choice over another is an illusion; actually we never had any choice in the matter. Even the sensation that we just made a free, unconstrained choice was predestined by fate. Or so say the determinists.

Thinking back to the Plinko analogy, we are like a chip that has fallen through many rows of pegs. We can't see other pathways that our world might have taken; for all we know, as judged from within our inescapable frog's-eye view of existence (page 153), there is only one path, and we are being channeled toward our ultimate fate. That is the deterministic worldview. But if there are many possible paths — which we can envision by taking a bird's-eye view where the Plinko pathways are constant-ly splitting — then the world might be indeterministic, free will might be recovered, and the other hypothetical paths might rep-resent the alternate realities and alternate universes of the many-worlds interpretation (page 118).

Recall that in John Wheeler's vision of it-from-bit physics, the bits that make up the world arise out of what he called acts of observer-participancy. According to Wheeler, we query nat-ure by seeking "the apparatus-elicited answers to yes or no questions." Observer-participants ask questions, either by setting up an experiment in a specific manner, by training a powerful telescope on an obscure patch of the Cosmos, or even just by looking over there rather than here. And nature responds with answers. In some cases, these answers may be entirely new infor-mation, their values derived at random: An experimenter may choose an axis along which to measure a particle's spin, and if the world has no information that constrains the result (i.e., nature doesn't already have the answer), then a random, indeterministic result will appear. It's then easy to see why any further decisions that the experimenter makes will be indeterministic as well, if those choices are in any way impacted by the result of the original measurement. Such further decisions don't have to be about the previously measured particle or the axis of spin measurements;

they don't even have to be about experiments. Chaotic influences, popularly known as the "butterfly effect," will ripple through and become amplified, affecting every further action the person is involved in, as well as all people and all things with which he or she comes into contact. The randomness of unconstrained measurements throws a twist into a world that might otherwise be deterministic, at least if everything consisted of stuff that strictly followed time-symmetric physical laws.

This provides us with an insight into the nature of free will. Earlier I mentioned the IBM computer Watson, whose performance on the TV game show *Jeopardy!* was fascinating to watch. With the help of a giant database of information, Watson had to take in each clue, interpret what was being asked, search the database for possible relevant responses, assign a confidence value to each possible response, and then choose one, based on this information about information. The process was algorithmic and not an example of free decision-making; one assumes that given the same clue and database, Watson would always respond identically. However, if only a few ingredients were added, Watson *would* display a kind of free will — at least, enough so that it might be indistinguishable from a human-type free will, for all practical purposes. For example, updating the database with each response, with Watson learning from its mistakes, would produce a dynamic where the same incorrect response is not given twice to the same clue. Biological memory faculties don't access all information equally at all times, so a random number generator could disable sectors of Watson's knowledge database for any given clue. Biological memories also copy information imperfectly, so Watson could be programmed to have a certain probability of "misspeaking" or "misremembering." Additionally, its performance could be modulated by its performance on recent clues, by increasing or decreasing the confidence values accordingly, thereby mimicking human emotional states where our confidence goes up or down depending on whether or not we're on a roll. Watson could be programmed to "get tired" deeper into the game, or to alter its risk-taking based on which contestant is in the lead, and by how much. Watson could even get distracted by sounds or by a competitor hitting a big Daily Double. It could have multiple, competing faculties with different agendas, for example one that wants to score big vs. one that doesn't want to "embarrass" itself, analogous to the Freudian id and superego, and the dominance of one faculty over another could vary with time and other game-related variables. In short, three factors — randomness, dynamics, and imperfection — could complexify the computer's decision-making to the point where it displayed a free-will-like quality. Of course, even with these multiple layers of complexity piled on, Watson's faculties would still be much, much simpler than a human's. People don't calculate numerical confidence values for each potential choice, for example; instead, it's an organic process involving images and language, and the emotional *and* intellectual feedback within the psyche that each one generates, and who knows what else.

I've also mentioned honeybees and the superorganism-like behavior that emerges when a colony swarms and leaves the hive to find a new home. In doing so, the swarm exhibits a kind of emergent free will that no individual bee exhibits. The decision-making even looks a bit like Watson's: Scout bees go off looking for a suitable location. If a scout finds one, it returns to the swarm and performs a variation of the waggle dance, which normally communicates the location of food sources. With multiple scouts returning and dancing, others go out to investigate the options. Gradually the balance tips to the option collectively decided upon as best, and when some critical point is reached, the entire swarm takes off together toward its new home. Groups of humans, too, exhibit similar decision-by-swarm processes. The global scientific endeavor is an example, with consensus on a topic gradually reached over years of experimenting and publishing, as well as criticism. Perhaps the most dramatic example of swarm-like decision-making is the democratic election of a leader. The process begins with many options that are slowly whittled down, perhaps to two. Through campaigning and debating, the candidates put themselves before the swarm, and on election day, the swarm decides one way or the other. The winner then becomes leader of the entire swarm, just as a new hive location becomes the home for an entire bee colony. Given the complexities, dynamics, and the fallible nature of the many millions of voters involved — not to mention the foibles of the candidates — the election process looks a lot like a blown-up, stretched-out version of one individual human, with billions of brain cells, trying to choose a toothpaste out of several dozen options.

With this treatment of free will, we begin to glimpse how *intentionality* or *purposeful action* comes out of a decision-making process. The more complex and convoluted is the informational chain between stimulus and response, the more the response looks like an intentional or purposeful act. If you touch a hydra and it contracts, or if you type 2 + 2 into a calculator and it comes back with a 4, it would be hard to call either example an act with purpose. However, if a deer hears what might or might not be a predator, weighs various options, and then decides to run, that looks more like what we humans would call intentional or purposeful. Like so many things in the informational view, intentionality is a function of complexity. Even a complex machine like a modified Watson might seem to be behaving with what we'd call "purpose."

God, Sickness & Death

From the introduction of this book, I've been upfront about looking for a way that the universe could appear to be fine-tuned for matter and life, without requiring a tuner. I've been an atheist pretty much my whole life. When I was about eight years old, my friend Danny Wong and I decided that the idea of a god who directs the course of events in the world is just silly. As I grew older, I learned more about science and the serious questions that theologians and others have raised about the nature of existence.

But I always thought that the desire for God-based explanations, even among the scientifically literate, arise out of an insufficient understanding of the world more than anything else. I figured that eventually, we would be able to explain things like fine-tuning naturalistically, the way we have explained the orbits of the planets as being undirected and self-sustaining. At the same time, I didn't like the necessity for trillions of unobservable alternate universes made of defined stuff, as a needed assumption just so the anthropic principle could produce that naturalistic explanation. There had to be a simpler way.

For me, Robert Lanza provided the bridge to a simpler approach, by floating the unspeakable conjecture that the universe is created by life — not the other way around. Even without countless ontologically existing alternate universes, and even without a designer god, if the act of observation gradually sharpens the laws of the Cosmos, we should not be surprised that after billions of years, the laws would naturally end up looking as if they were fashioned explicitly *for* life. This is extremely attractive in light of Occam's razor, since we are stripping away trillions of unseen but ontic universes (deemed necessary by the anthropic principle to account for observations), as well as a complex, intelligent designer whose own origins and properties are completely mysterious. The question remained, though, how a life-driven universe could be accomplished. This was especially challenging given the assumption, based on airtight science, that life is made of the pre-existing matter of the universe. Lanza's proposition seemed to be absurd, scientifically a non-starter. But there was something that seemed *just so right* about it, and I refused to dismiss it on sight.

After following Lanza to the work of John Wheeler, and then discovering Carlo Rovelli's relational interpretation of quantum mechanics, the pieces began to come together. Rovelli suggested that the state of any object is, in a literal sense, a relation with some observing system; this state is not an absolute, observer-independent, and enduring property possessed by the object, and therefore by the world in general. A breakthrough came in realizing that informational context sharpens other information to produce more precise and detailed descriptions of the world. By realizing that contextually constraining information may have been been accumulating for billions of years, it became possible to extrapolate backward to an extremely simple informational beginning — far simpler than the material beginning of the Big Bang. The final piece came when I realized that all organisms are causally linked and are therefore effectively subsystems of a single observing system. The simplest-case scenario then became a complete account for our observations, including the appearance of fine-tuning. It is an explanation for the world arising out of the fewest elements or building blocks possible. The entire universe emerges from a unified structure composed of only one ingredient, the fundamental informational relation, or the bit. No designer god is required.

Scientists can be defensive about anything even approaching religious themes; consider the reaction to Paul Davies' op-ed article about faith in science (page 145). Personally, I believe this is due to a general unease with the idea of a naturalistic universe made of stuff. Every stuff-based explanation has difficulties, and the proposed solutions often conflict with Occam's razor and/or the tenet of testability in science — and scientists realize this. Thomas Nagel, in *Mind and Cosmos*, tries a radical approach by supposing some kind of teleological (purposeful or directed) principle driving the course of events in the world, even if that impetus isn't necessarily a creator god. This demonstrates the desperate measures necessary in these desperate times. But what if this impetus is only an illusion, conjured in retrospect by conscious and self-aware observers, for whom a detailed and vast universe seems to have been fine-tuned for the purpose of producing them? Quoting Davies one last time, "Here we have a universe that has an explanation within itself: The observers that arise, play a part in selecting the very laws that lead to the emergence of observers in the first place." There is an evolution in the complexity of observing subsystems; the progressively more complex sub-systems then sharpen the universe's observed properties and laws, a process of top-down causation — which also mirrors the top-down causation that happens within life itself (page 201). This provides a mechanism for the top-down cosmology pro-posed by Hawking and Hertog (page 177). The longstanding difficulties with the traditional reductionist/bottom-up approach, which Nagel and others have articulated, are wiped out by the top-down approach.

I don't wish to disappoint anyone, but if we can account for observations without invoking a designer god, we must also razor away an unobservable heaven, some place where we go when we die. What is commonly called the soul is a phen-omenon of consciousness, emerging from both the decades-spanning continuity of one's identity and the active exchanges of information in the mind. A major clue into the nature of "the soul" can be seen in what happens when we are sick. The mind is hardly unmolested by physiological illness. When you're in bed with a high fever, your mind becomes delirious. It's in a bath of information about excessive environmental temperature, caused indirectly by an invasion of other subsystems that we call bac-teria or viruses. And, because biological information-maintenance and copying is imperfect and is always being challenged by new information (as in the case of everything from mutations to asteroid impacts), when you have a fever, you cannot process information the way you can when you are healthy. The sensation of there being a soul emerges from healthy persons' informa-tional processes and memory faculties, which together make up consciousness. In sick people, those processes can be interfered with; we've all witnessed this as well as experienced it personally. Information-processing abilities can be damaged permanently. If the brain suffers a physical trauma — which, we remember, is an emergent stuff-based way of describing things — or in old age, information-processing functions can be lost entirely. It's all about information, and information can be imperfectly reproduced as well as destroyed. That's not compatible with the idea of a soul that survives death and lives for an eternity in some perfect realm and/or perfect state.

So, what is death about then? Death is the loss of any subsystem as a potential processor or reproducer of information. I mean that broadly, because there are many kinds of subsystems. Even if a person dies suddenly, there are cellular processes that continue going on for minutes to hours. In the case of something like a heart attack, the subsystems first to be lost may be those associated with molecular processes in heart muscles that require oxygen to continue. The cardiac muscle cells subsequently lose their ability to convert chemical energy into the mechanical energy of contraction. Causality continues in this bottom-up fashion until the heart, as an organ-level subsystem, goes into fibrillation-type patterns, and the blood stops circulating. When that happens, the direction of causality changes to *top-down*: Circulatory changes cause brain cells to respond to information on low oxygen levels, and their molecular-level information processing ceases, too. Brain cells die — those subsystems are functionally lost — and as a result, the highest functions, what we call consciousness, cease to operate. Top-down causation continues until all of the molecular-level informational processing halts, permanently. Notice that even in dying, there is bidirectional causation flow. But all of that is over when the person is fully dead: The cold, dead corpse is not processing any information whatsoever. That isn't counting the bacteria and other species left behind, of course, many of which are just getting started processing the information remaining about their host.

Surely, a person's mental functions depend partly upon the informational network of the mind, corresponding in the stuff approach to the configuration of neurons and synapses — the "wiring" of the brain. But they also depend upon the second-by-second dynamical aspects: the informational pathways that are active at any given moment, corresponding to electrochemical potentials traveling along neurons. When you are dead, all of the latter activity is over. Meanwhile, in a dead person, the street-map information representing the spatial configuration of the neurons and everything else begins to get lost, as well. The tissue starts to turn to mush, and if you examined the chemistry of a deteriorating cell, you would discover that the unchecked action of enzymes is largely responsible, a process known as autolysis. Autolysis occurs in the same manner that all non-living configurational information tends to degrade over time, due to the inevitable increase in entropy (pages 66 and 200). Gradually, all of the information that made up the person is dispatched to thermodynamic oblivion — the bit-based equivalent of ashes and dust.

Viewed from an informational perspective, death and decay are more profound than they are from a stuff perspective, where the distinction between life and non-life is only a matter of the coordination of molecules interacting. Informationally, death means the irreversible loss of the relations that defined the sub-system and therefore the person. Experiments investigating differences between living and dead microorganisms could shed light on this distinction as well as the physical nature of life itself (see the Appendix).

There can be no life after death because there is no dynamic, bidirectional life-information after death. There is only matter-information, and like all matter-information that isn't maintained and repaired by the entropy-reversing processes of life, that degrades. However, as many people have speculated, someday we may be able to upload our mind and conscious-ness to some form of technology, like the way we can back up a hard drive. I have no idea how this might be accomplished, and it might prove extraordinarily difficult, but in principle, it is possible. Taking seriously the idea that informational inter-actions on the molecular level represent the building blocks of life and consciousness, we could start by creating computer simulations of these interactions from a very simple organism, such as a genetically minimal bacterium. The ultimate goal is extremely attractive: If we could reproduce ourselves by up-loading our conscious identity to something like a quantum computer, not only would the technology give us immortality, but the possibilities for human–technology hybridization and advancement would be endless. This would certainly be a giant leap forward for mankind, placing us squarely in a Stage 8 system: the techno-dominant universe.

The ultimate destination for humans could be that our biological bodies — squishy structures made of warm meat — become obsolete. This will be achieved when technology can perform all of the information processing that biology can do, at all levels, only more efficiently and less prone to error. Science fiction notwithstanding, we won't be brains in a vat, and we won't be suspended in pods wired up to the internet. The universe began with information that evolved, and that's what will drive our future technology-based evolution as well.

Flip a Switch, Change the World

You may not be able to achieve technological immortality today, but the universe isn't about you. It's about *us* — and collectively as a super-observing superorganism, we have had a kind of immortality since the very beginning. Every observation that's been made, by every organism that has ever lived, has gone into the melting pot of information that makes up the universe. Any time you flip a switch, or throw a rock, or detect a particle, you are changing the world. Thereafter, the world becomes constrained by your actions or your discoveries, forever. We often say similar platitudes when a loved one dies — "She will live forever in our memories" — but in the simplest-case scenario, that kind of immortality is a physical fact of nature. As an observing subsystem, you are a creator-god of the universe, one of the latest in a very long line. And even though the orchestra of your personal consciousness may go quiet someday, your informational legacy in the world will remain — just like the memories you leave in the lives of everyone you touched.

"Can we ever expect to understand existence?" Asked John Wheeler. If we think of the world as a collection of independent particles that we passively look upon, as isolated individuals, it's no wonder the universe seems cold, lonely, and pointless. In the future, this worldview will be vanquished to history, like the ancient belief that our lives are at the mercy of capricious, judgmental gods. The informational view, which we're only starting to understand, reveals nothing short of the Meaning of Life — and when he came up with "it from bit," that's precisely what John Wheeler was chasing. "If and when we learn how to combine bits in fantastically large numbers to obtain what we call existence," he wrote, "we will know better what we mean both by bit and by existence."

When we figure out how information underlies everything, we will finally have the tools to take the next technological and evolutionary steps. Only then will we be able to transform our civilization into Carl Sagan's ultimate ideal: We Who Became One.

QUESTIONS & ANSWERS ABOUT THE SIMPLEST-CASE SCENARIO

How could you have possibly cracked the code on existence, when thousands of highly trained physicists, philosophers, and biologists before you haven't?

I may have gotten very lucky. I was in the fortunate position of being able to read physics papers online at my leisure, for years — being at the right place in life, at the right time. Before the internet, this kind of vigorous research would have been much harder. Also, career physicists tend to be highly specialized; the vast majority focus on what Thomas Kuhn calls "ordinary science," which proceeds in careful, incremental steps. Physicists who are able to tackle big, foundational issues ("revolutionary science") are few, because there just isn't much funding for that kind of work. Perhaps it took a dedicated amateur, by sheer luck, to stumble upon the pieces and put them together just so. That said, if the simplest-case scenario has any merit, it will take professionals across various fields to formulate a rigorous and workable theory. At best, I may have just gotten the ball rolling.

According to the philosopher of science Karl Popper, a scientific hypothesis or conjecture must be falsifiable, or it isn't science. How could the simplest-case scenario be falsified?

Finding or receiving information from a distant civilization with its own detailed history of observing the same "sharp" universe would prove that the positions of stars, for example, are not exclusive to our local observations — it would show that the stars are ontologically independent of our lineage of life. This would be analogous, in the theory of biological evolution, to finding a rabbit fossil in the precambrian layer of the Earth, which was laid down well before the evolution of mammals. No such thing has ever been found, and evolution having stood up to falsification in this manner (among many other reasons), it is considered to be a well-tested scientific theory.

The absence of evidence for aliens is not evidence of absence.

Until such time that we observe alien life, or something with similar complexity, there are two possibilities: (1) Nothing is observed and aliens are *not* present, or (2) nothing is observed but aliens *are* present. The greater the confidence with which we can rule out possibility (2), the greater the confidence we can place in (1). The kind of massive spectroscopic scanning program described hypothetically in the Introduction, for example, could make the second possibility increasingly less likely, and therefore make the first possibility increasingly more likely.

So Earth, and we humans, are at the center of the universe?

Not at all. We'd really only be at the center if there were an absolute framework of space, marked off with a cosmic ruler with a zero value at our home planet, and there is no such thing. Even in the mainstream tradition of stuff, our position in space is relational. Look at it this way: If you found yourself in a large desert, and you saw a circular horizon all around you, is this evidence that you're at the center of the desert? If you walked a long distance, you'd probably still see the same thing. This would disprove your earlier interpretation that you were at the center.

How is it not arrogant to claim that we're the only life in the universe?

There's no evidence of other life, and it's not too arrogant to make a provisional claim that's consistent with observations. The arrogant position is to claim we are quite sure of life being out there, because we are accustomed to thinking of the world as being made of stuff. That's especially problematic when the conventional stuff approach clashes with experiments. Isn't it more arrogant to claim that the universe is an extremely vast place full of defined particles, virtually 100% of which we will never observe or detect, merely because it appears to us to be set up that way? Arrogance is insisting that that's the only valid interpretation, even when a far simpler interpretation may be at hand.

I don't want to believe that we're alone in the universe. Are there any ways in which there could be life on other planets?

Three ways. First, there may be life elsewhere in our Solar System, descended from the same original lineage as Earthly life. Second, we may be in a panspermia situation, where life (also derived from a common ancestor) has been seeded across the galaxy, although the distances involved make this conjecture controversial. Third, the simplest-case scenario may admit the possibility of *very simple* lineages of life arising, seemingly independently, in our universe. The stumbling block for finding truly alien life is that the history of any other informational systems needs to be consistent with our techno-biological history. If they were Stage 1 or maybe Stage 2 systems (pages 236–239) — for example, not distinguishing between directions corresponding to the three spatial dimensions that we observe — then perhaps such another extremely simple history would be compatible with ours. This would be the informational equivalent of universes colliding. However, the possibility leaves open little room for alien intelligence, or significant complexity. I want to guess that statistically, we likely wouldn't find systems Stage 4 or higher by searching anywhere in the observable universe. But I don't know.

Are there any experiments that could test the simplest-case scenario?

I have an experiment proposal that would demonstrate that nonlocal "spooky action at a distance" comes from entanglement between techno-biological observers, not between entangled particles. A small object made of a radioactive material is placed in a superposition "cat state" within a chamber. Then, two portholes are opened, allowing the decay products to leave the chamber. Detectors at either porthole measure the particles' momenta, and from this information, the object's classical configuration is determined. The prediction is that both detectors will always find the same configuration. In the "stuff" tradition, those results would suggest that the object somehow knows when the portholes are being opened, and collapses into a classical state, even before enough time has elapsed for any porthole information to traverse the spatial distance back to the object. Or, when the portholes are opened, something seemingly goes back in time to tell the object that it's about to be observed, and that the object must now take a distinct, classical configuration. Otherwise, how does the object "know" to emit particles with momentum information corresponding to one configuration or the other? The simplest-case scenario says that it doesn't: The detectors, being mutually constrained (entangled) techno-biological observers, are simply registering consistent information. The particles are not ontically leaving the object at any time; particle-information does not appear in the world until registered by the detectors, which are nonlocally constrained such that this information is always correlated. This is like an entanglement experiment, finding similar nonlocal correlations across space, except that no entangled particles are prepared — thus indicating that it's the *observers* that are entangled.

Other experiments could look for fundamental differences between functioning and non-functioning techno-biological observers. For example, the simplest-case scenario predicts that two functioning nanobots, mutually measuring each other, could not be placed into a superposition cat state. If the nanobots are registering information, they will mutually collapse their partner's subsystem's superposition. However, as long as the nanobots are disabled or nonoperative, the superposition can be maintained. Still other experiments could look for subtle changes in the quantum dynamics of microorganisms when they are killed by the zap of a laser.

Can the simplest-case scenario lead to anything that's actually useful or practical, such as a new technology?

If the program reflects the fundamental nature of the world, definitely. But I'm not smart enough to invent such new technologies; it's impossible to predict ultimately what will come out of any scientific revolution before it happens. Who could have predicted the invention of the oxygen tank, or the CO_2 fire extinguisher, before we learned about the different gases?

The simplest-case scenario sounds like a newer expression of the *idealism* school of philosophy, that physical reality is in some sense mental rather than material. Samuel Johnson's famous response to idealism was to kick a rock and declare, "I refute it thus!" How do you refute Samuel Johnson?

Samuel Johnson did not know that the atoms in the rock and in his shoe are actually 99.9999% empty space, or that his foot bouncing off the rock was due to electrostatic repulsion rather than "solidity" (whatever that may mean). For eons, we biological organisms had observed that solid objects cannot pass through each other, and the universe being a logically consistent place that appears to follow physical laws, when Johnson observed a rock, and imparted momentum information into his foot, naturally we (and he) should expect to observe his foot to bounce back. But this does not prove the materialism of the foot or rock, any more than performing this same act in a well-rendered virtual-reality game would demonstrate that your character's foot, or an onscreen rock, are made of material particles. Show me a sufficiently convincing video game, and I will refute Samuel Johnson thus.

The appearance of a collapse of the wavefunction is often explained by *decoherence* theory. How does the simplest-case scenario treat this explanation?

Decoherence is a brilliant discovery — a mechanism that is routinely demonstrated in the lab, wherein quantum superposition is "leaked" into the environment by an object's interaction with that environment. Physicists are able to put small objects into a superposition of multiple states. However, if the temperature is allowed to rise more than a few thousandths of a degree above absolute zero, then infrared photons enter the picture, and these environmental photons are said to interact with the system that's been placed in superposition, effectively observing it. When this happens, the superposition seems to disappear, and the system is seen to take a classical state. Thus, decoherence theory has shown why quantum superpositions are not ordinarily observed in objects of any size. Decoherence explains, through an experimentally demonstrable mechanism, why we never see a baseball in two places at once, for example.

There are some subtleties at work here, however. In decoherence experiments, there is a clearly delineated boundary between the object and the environment, and human experimenters create these conditions. But in other situations, even according to the "stuff" approach, there is no such boundary. If you want to impress a physicist familiar with decoherence, ask how the early universe can be said to have undergone decoherence into classical structures, when there was no delineation or distinction between objects and observers at the time. In that closed system, what accounts for the partitioning necessary for objects to be observed by their environment? It's an unresolved problem. Also, interpreting decoherence experiments under Carlo Rovelli's relational quantum mechanics, where the state of the object can only be considered a relation rather than an absolute state, the experiments are less cut-and-dried than they otherwise seem; RQM makes us examine the very nature of superposition.

Being a stuff-oriented mechanism, it may be that decoherence emerges out of information just like every other mechanism or law of nature. However, if we allow that observers, fundamentally speaking, are subsystems of the techno-biological superorganism, with objects emerging in relation to these subsystems, then informational mechanics at least gives us a way to set the boundary between objects and observers. This boundary is sometimes called the *Heisenberg cut*, and in its connection to an ongoing dilemma called the *measurement problem*, physicists have debated for decades where exactly the Heisenberg cut should be placed.

How would the simplest-case scenario respond to Einstein when he objected to quantum mechanics by asking whether the Moon is there when you aren't looking at it?

This question hinges on what we mean by the word "there." Does "there" mean that all of the Moon's particles continue an ontologically defined existence across time while un-observed? If so, the answer is no. In the simplest-case scenario, "there" in this context might mean something like, "very likely to be observed in that spot," or, represented by a probability function that approaches 100% certainty, based upon past observations. In that understanding, the Moon is always "there," no matter who is or isn't looking at it.

There's do-it-yourself demonstration where you can build a cloud chamber in an empty aquarium, and you can see nuclear-decay particles making tracks through the air. When you aren't looking, what's going on in the tank?

The first thing one needs to ask in such a case is: How did that demonstration come together? It was assembled by humans. Someone had to buy dry ice and put it into a closed space with alcohol and whatnot. Science has discovered the element radon, and it has discovered that radon is a trace component of the atmosphere, and that a radon nucleus gives off a particle when it decays, and that such particles create visible tracks in cold alcohol vapor. Given these discoveries in the mainstream science of stuff, when we create conditions in which these laws of nature tell us that particles ought to be observed, then we should not be surprised to observe the particles. However, a human creating these conditions of constraint, and then registering appropriately constrained information that conforms to the discovered laws of nature, has no bearing on the ontological status of the stuff corresponding to this information. This is a long way of saying that if we see tracks in our homemade cloud chamber, we cannot assume that those same tracks fill the air around us all day long in an ontic sense, the only difference being their visibility.

The direct answer to this question is: I don't know. However, I suspect that any techno-biological observer, including a home-made cloud chamber (similar in many ways to a manmade photographic plate), is capable of bringing such information into the world. A similar thought experiment: When did information about the configuration of the far side of the Moon first appear? No living or technological thing had ever observed its features until the Soviet Union sent an unmanned spacecraft around the back in 1959. Did that information first become extant when the spacecraft's camera photographed the far side? The simplest-case scenario would say yes. Since it does not consider biological, conscious, or human observers special compared to technological observers, a manufactured camera would do the trick.

If the structure of molecules is a human discovery, how can an insect's sensory organ detect a single molecule?

Quantum mechanics shows us that information comes in discrete, irreducible packages. The photon is the iconic example, but all stuff-information is quantized in a similar manner. So, for instance, we humans might describe an ant pheromone as coming in packages called molecules, and these packages are indivisible in a sense, because half of a molecule isn't the same pheromone anymore. But in relation to a sensory organ in an ant's antenna — a very simple informational subsystem — what we call a molecule is a discrete, indivisible package of information which, relative to the ant's organ, might be called "A." If the sensory organ had evolved to detect A, then the registration of "yes" information provides the ant with a message that's something like, "yes with regard to the presence of A." This is a sharpened form of information that the organism can then use to its benefit. So, it's not that the ant is detecting a molecule, per se. It's registering a package of information. We humans, with our scientific knowledge and lab equipment, are the ones that interpret the same information with much sharper detail, including talk about atoms of hydrogen and carbon and the angles between the chemical bonds, for example.

You claimed (Chapter 8) that life or technology is required for information to "stick" in the world. If something hits the Moon and makes a crater, isn't that an example of information persisting long-term?

It may seem so, because a crater on the Moon lasts for a long time. But if you could speed up the movie of change to an extreme degree, you'd see the crater eroded away into nothingness eventually, or covered by other craters. This is is what eventually happens to every physical feature of the world whose integrity is not maintained by the local-entropy-reducing processes of life or technology. In contrast, life uses information-destruction (mutations) to its own advantage — one of life's unique features.

Regardless of your arguments, existence cannot dream itself into existence.

There are many ways to phrase objections to the simplest-case scenario that make the proposal sound absurd. This one is a straw-man portrayal; nothing is "dreamed into existence." If the universe is a closed system, then one can argue (as Paul Davies has; see pages 182–183) that existence can only arise from within that system: Something can arise within a closed system only if the system brings that something into existence. Therefore, if we're talking about the *beginning* of a closed system, with things in that system arising out of nothing, as it were, then that system in some sense has to *bring itself into existence*, with no outside help — a little like how the debts among the four friends on page 105 are brought about only by the friends themselves. Such a beginning may seem astonishing, but the universe doesn't care about our misinterpretations or our astonishment.

AN INFORMATIONAL GLOSSARY

Semi-rigorous definitions for common terms as used in this book. Words not included are assumed to have ordinary and intuitive definitions. I have done my best to avoid circularity, but keep in mind that in a relational world with no absolutes, everything is necessarily defined in terms of other things. See the discussion about the definition of the second on page 74, for example.

Alien: Two **subsystems** are alien with respect to each other if they share no common causal origin, and therefore are subsets of different, topologically disconnected **systems**.

Bit: The elementary **informational relation**, which has two possible varieties, designated as either 0 or 1 in digital technology. The labels we put on the bits are arbitrary.

Consciousness: A unified experience of the **world** that **emerges** to complex **subsystems**.

Constraint: A condition in which **information** appears to a **subsystem** not with random **values**, but rather, with a reduced range of values or only a single value, as a result of **observations** by other subsystems and the internal logical consistency of the **world**.

Contemporaneous description: A **description** of a **thing** or **event**, relative to a **subsystem**, based on minimal **information** available to that subsystem. See **Modern description**.

Death: The irreversible loss of a **subsystem** as a potential recorder, transmitter/**reproducer**, or processor of **information**.

Description: A complexified collection of **information** that embeds **meaning**, for example by including a number representing a ratio with a standard **unit**.

Distance: The quantification of **separation** by **space**, or a number of **units** expressing that separation.

Duration: The quantification of **separation** by **time**, or a number of **units** expressing that separation.

Emergence: The appearance of complexity arising out of a less-complex underlying structure. **Things** such as **matter** and **energy** emerge out of the **information** of which the **world** is **fundamentally** made.

Energy: A type of **stuff** that **emerges** from **information**. The energy possessed by an **object** is not **absolute**; it depends upon the **reference frame** of the **observer**. The term can also refer to a **quantity**, analogous to **mass** as it is commonly used to describe a relative quantity of **matter**.

Entanglement: A phenomenon that **emerges** as a result of the **constraint** of **subsystems**. The entanglement of two **particles**, for example, emerges out of the constraint between the subsystems making **measurements** of them.

Event: Any point in **spacetime** that can be defined in relation to some **thing**, by using a **description** that specifies a particular **reference frame**.

Evolution: Broadly, the progression of a **system** through different configurations, generally increasing in complexity. Specifically, the term can refer to individual species progressing in a similar manner, although not necessarily increasing in complexity.

Extension: Deviation from a point-like **event**; extension may be expressed as a line segment in **spacetime**. If a **thing** or process has the property of extension, then parts of it are **separated** from other parts, either in **space** or in **time**, or both.

Free will: The ability of a **subsystem** to select from among two or more possible choices, in a manner that is not **constrained** and that cannot be predicted by other subsystems via their **knowledge**.

Fundamental: Irreducible; not derived from or based any other entity. **Information** is the fundamental ontological component of the universe, out of which **things** appear via **emergence**.

Future: The direction along the **emergent** continuum of **time** from which new **information** never appears to an **observer**. By definition, information always appears from the opposite direction, the **past**.

Information: Any quantity, number, or **description** that reduces uncertainty about some aspect of the **world** for some **subsystem**.

Informational relation: The form that **information** takes in the **world**. Information is never **absolute**; information can only exist or can only have **value** as considered **relative to** some **subsystem**. Therefore, to gain information is to acquire an informational relation.

Knowledge: The **information** that is specific to, and (in principle) accessible by, a **subsystem**. In cases where this information contradicts **objective reality**, the knowledge is considered to be false.

Life: A quality that marks the ability to make **observations** and to store, **reproduce**, or otherwise process the resulting **information**, as a biological (i.e., non-technological) function.

Mass: A quantity of **matter** (or its equivalent **energy**) that **emerges** from the perspective of a **subsystem**, given a particular **reference frame**.

Matter: A type of **stuff** that **emerges** from **information**. The mass associated with an **object** is not **absolute**; it depends upon the **reference frame** of the **observer**.

Meaning: A complexification of **information**, such that it is capable of reducing uncertainty beyond the either–or, on–off **values** of a single **bit**. Information that embeds meaning is a **description**. In this book, a more meaningful description is sometimes called a more detailed or **sharper** description.

Measurement: A comparison between two or more **observations** (one of which may be the establishment of the **reference frame**) that is expressed in terms of some standard **unit** of measurement, often as a numerical ratio.

Modern description: A **description** of a **thing** or event, relative to an **observer**, incorporating contextual **information** that increases the description's **sharpness** or **meaning**. Typically this contextual information arrives to the observer later in **time**, such that the modern description becomes a more meaningful description of an event in the **past**, i.e., as a relation across time that's more detailed and informationally rich than the original or **contemporaneous description**.

Object: A **thing** composed of **matter** or **energy** in **particle** form, although more generally, the term might also include mathematical objects such as equations.

Objective reality: The configuration of the **information** in the **world** that **constrains** the appearance of new information. Objective reality may differ from **knowledge**.

Observation: The registration of **information**, or the appearance of a new **informational relation**, from the perspective of (i.e., **relative to**) an **observer**.

Observer: Any biological or technological **subsystem** with the potential to have **information** or engage in **informational relations**. Just as it is routinely accepted that velocity can only be described **relative to** some **thing**, information only exists relative to some specified observer, and only has **meaning** in relation to or relative to the observer.

Particle: The form in which a very small **thing** emerges relative to a **subsystem**, in the case that specific position or momentum **information** on that thing exists in the **world**.

Past: The direction along the **emergent** continuum of **time** from which new **information** appears to an **observer**.

Photon: A **particle** of electromagnetic **energy** (for example light or radio waves), or its **wave** equivalent.

Physics: The study of the **world** as it relates to the behavior of **things**. Physics seeks to understand this behavior in terms of the simplest laws governing the simplest things or processes. Most generally, physics is the study of the interaction of **information** in the world.

Present: The zero-**duration** boundary between the **past** and the **future**; a phenomenon of **consciousness**.

Reality: See **Objective reality**.

Reference frame: A specific system of coordinates in which a **measurement** is made, necessary for the measurement to have

meaning and therefore serve as a non-arbitrary **description**. The reference frame provides the basis for the measurement. In a three-dimensional world, it typically includes a location in **spacetime** for the origin (or zero point), an orientation of the three spatial axes, a **unit** of scale, and a chirality ("handedness" convention).

Registration: The storage or fixing of **information**, either new or **reproduced**, in some durable form. **Things** are seen to **emerge** out of information that has been registered.

Relation: See **Informational relation**.

Relative to: In relation to; as seen from the perspective of; as would be described from the viewpoint of; when compared to. Example: A jet airliner takes off approximately 150 miles per hour relative to the runway.

Reproduction: Any copying or transmission of **information** from one **subsystem** to another.

Sentience: The ability of a **subsystem** (1) to send or receive **information** from other subsystems (or within its own subsystems) in the form of complexified symbols that embed **meaning**, (2) to be aware that it is doing so, and (3) to have **free will** over doing so. Humans are sentient because we speak, write, and think using words and sentences, we understand that we are doing this, and we can do this however and whenever we choose.

Separation: Two **things** are separated if they do not share the same location in **spacetime**.

Sharpness: The quantity of **meaning** or **informational** detail embedded in a **description**. The **unit** for this measure is, ultimately, the **bit**.

Space: A type of **separation** that **emerges** between different **subsystems**. Space can be quantified as **distance**.

Spacetime: The substrate that **emerges** from the perspective of **life**, combining both **time** and **space**, in which **matter** and **energy** are **observed** to follow laws of **physics**.

Stuff: The material that makes up the complex **world** that **emerges**

as seen from the perspective of **subsystems**, consisting of **matter** and **energy**, which follow laws of **physics** as they move and interact in **spacetime**.

Subsystem: Any subset of a **system** containing at least one **informational relation**.

System: A closed, self-contained configuration of **informational relations** that exhibits internal logical consistency and therefore is subject to **constraint**.

Thing: Any **object, particle, wave**, etc., that **emerges** relative to a **subsystem**, out of the **information** accessible by that subsystem.

Time: A type of **separation** that **emerges** out of changes to a **subsystem** or a **system** as a whole. Time can be quantified as **duration**.

Unit: A standardized **measurement** against which other measurements can be compared, often in the form of a ratio; for example, the **description** "ten seconds" is a **duration** consisting of ten second-units of duration in succession.

Universe: A **system** of **information**, as perceived or observed by some **subsystem** of that system. Synonymous with **world**, although typically used to describe every **thing** that exists in the world.

Value: A number, or a range of numbers, associated with **information**. For example, a **bit** in a computer system can take a value of 0 or 1. A **description** adds **meaning** to a value; for example, in a description, a value can take the form of a rational number associated with a standard **unit**.

Wave: The form in which a **thing** emerges relative to a **subsystem**, in the case that no specific position or momentum **information** on that thing exists in the **world**.

World: A **system** of **information**, as perceived or observed by some **subsystem** of that system. Synonymous with **universe**, although it's typically used to describe the perceivable external and internal environment that **emerges** in relation to a subsystem.

BIBLIOGRAPHY

Karl Coryat, Toward an informational mechanics, entry in the Foundational Questions Institute's essay competition "Questioning the Foundations" (2012).

Paul Davies, Taking science on faith, *The New York Times*, November 24, 2007.

Paul Davies, interviewed by Robert Lawrence Kuhn on PBS television series *Closer to Truth*, episode 912 (2011).

Terence Deacon, What is missing from theories of information?, in *Information and the Nature of Reality*, P.C.W. Davies & N.H. Gregersen, eds., Cambridge University Press (2010).

Daniel Dennett, *Darwin's Dangerous Idea*, Simon & Schuster (1996).

George F.R. Ellis, *Top-Down Causation*, G.F.R. Ellis, D. Noble & T. O'Connor (eds.), Interface Focus **2**(1) (2012).

Lewis Carroll Epstein, *Relativity Visualized*, Insight Press (1985).

Hugh Everett III, The theory of the universal wave function (1956), Princeton University dissertation.

Richard Feynman, Robert B. Leighton & Matthew Sands, *The Feynman Lectures on Physics, Volume 3*, Addison Wesley (1971).

Stephen Hawking & Thomas Hertog, Populating the landscape: a top down approach, Physical Review **D73**(12:123527) (2006).

Werner Heisenberg, *Physics and Philosophy*, Penguin (1958).

Lawrence Krauss, *A Universe From Nothing*, Atria Books (2012).

Lawrence Krauss, The energy of empty space that isn't zero, Edge.org (2006).

Thomas Kuhn, *The Structure of Scientific Revolutions*, University of Chicago Press (1962).

Bernd-Olaf Küppers, Information and communication in living matter, in *Information and the Nature of Reality*, P.C.W. Davies & N.H. Gregersen, eds., Cambridge University Press (2010).

Robert Lanza, A new theory of the universe, *American Scholar*, Spring 2007.

Robert Lanza, *Biocentrism*, BenBella Press (2009).

John Marburger, On the Copenhagen interpretation of quantum mechanics, Symposium on "The Copenhagen Interpretation: Science and History on Stage" (2002).

Thomas Nagel, *Mind and Cosmos*, Oxford University Press (2012).

Carlo Rovelli, Relational quantum mechanics, International Journal of Theoretical Physics **35**: 1637 (1996).

Carl Sagan, *Cosmos*, Random House (1980).

Claude Shannon, A mathematical theory of communication, Bell System Technical Journal **27**: 379–423 & 623–656 (1948).

Lee Smolin, *The Trouble With Physics*, Houghton Mifflin (2006).

Henry Stapp, *Mindful Universe*, Springer (2007).

Max Tegmark, Parallel universes, in *Science and Ultimate Reality: Quantum Theory, Cosmology, and Complexity*, J.D. Barrow, P.C.W. Davies, C.L. Harper Jr. (eds.), Cambridge University Press (2004).

Vlatko Vedral, interviewed by Paul Davies on www.amazon.com (on page for *Decoding Reality: The Universe as Quantum Information*).

Sara Imari Walker & Paul Davies, The algorithmic origins of life, Journal of the Royal Society Interface **10**(79):20120869.

Sara Imari Walker, Is life fundamental?, entry in the Foundational Questions Institute's essay competition "Questioning the Foundations" (2012).

John Archibald Wheeler, Information, physics, quantum: the search for links, in *Complexity, Entropy & the Physics of Information*, W. H. Zurek, ed., Westview (1990).

Norman Wildberger, *Divine Proportions: Rational Trigonometry to Universal Geometry*, Wild Egg Books (2005).

David Wiltshire, Cosmic clocks, cosmic variance and cosmic averages, New Journal of Physics **9** (October 2007).

H. Dieter Zeh, Quantum discreteness is an illusion, arxiv: 0809.2904v5 (2009).

ACKNOWLEDGMENTS

This book never would have been conceived without the tantalizing insights of Robert Lanza, which include connecting Hawking's top-down cosmology to the idea that the Big Bang is what we would see if we could witness the birth of the universe ourselves. Carlo Rovelli's work was critical as well; I'm quite certain I couldn't have expressed the simplest-case scenario with any coherence were it not for his relational revelations. And even though his subsystem was dispatched to thermodynamic oblivion in 2008, I have to thank John Archibald Wheeler, whose landmark "it from bit" essay literally contains clues to how the universe is put together. I wish I could have had a conversation with him.

The Foundational Questions Institute (fqxi.org) served as the platform on which I assembled my ideas. There is no other place where a rank amateur can put forth a radical scientific proposal, and if it's interesting and well expressed, world-class scientists will read it and even comment. The fact that sizable cash prizes are given away, too, is almost unbelievable. Special thanks to Zeeya Merali, Brendan Foster, Max Tegmark, and Anthony Aguirre for staying in touch with the people.

Russell McDonald, Steven Clark, and Michael Aparicio read an early version of the book and gave invaluable feedback. Thank you for volunteering your time on such a long-shot endeavor!

The participants at the online Physics Forums have always been helpful in answering my questions and pointing me toward good resources.

Finally, I salute everyone who asks the biggest questions in the universe, and especially those who attempt to find answers. We'll figure it out one of these days.

www.ingramcontent.com/pod-product-compliance
Lightning Source LLC
Chambersburg PA
CBHW021420170526
45164CB00001B/30